RETURN OR RENEW

JAN 3 0 2025

BY THIS DATE

YOU ARE RESPONSIBLE

for all items checked out on your record.

KNOW WHEN

your items are due.

AVOID CHARGES

by returning or renewing on time.

RENEW ONLINE

by logging into "My Account" at
library.umaine.edu

RENEW BY PHONE

by calling the Circulation Department
at 207-581-1666

REPLACEMENT COSTS

are typically $55.00 per item
damaged or not returned.

ALL MATERIALS ARE SUBJECT TO RECALL AFTER TWO WEEKS

1865 THE UNIVERSITY OF MAINE

Fogler Library

CHEMICAL NOMENCLATURE

A collection of papers comprising the Symposium on Chemical Nomenclature, presented before the Division of Chemical Literature at the 120th meeting—Diamond Jubilee—of the American Chemical Society, New York, N. Y., September 1951

Number eight of the Advances in Chemistry Series
Edited by the staff of *Industrial and Engineering Chemistry*

Published August 15, 1953, by
AMERICAN CHEMICAL SOCIETY
1155 Sixteenth Street, N.W.
Washington, D. C.

Copyright 1953 by
AMERICAN CHEMICAL SOCIETY

All Rights Reserved

CONTENTS

Introduction 1
    Austin M. Patterson, 221 North King St., Xenia, Ohio

Letter of Greeting 3
    A. F. Holleman, University of Amsterdam, Amsterdam, Holland

Some General Principles of Inorganic Chemical Nomenclature 5
    Henry Bassett, Geological Survey Department, Dodoma, Tanganyika, Africa

Nomenclature of Coordination Compounds and Its Relation to General Inorganic Nomenclature 9
    W. Conard Fernelius, The Pennsylvania State College, State College, Pa.

Problems of an International Chemical Nomenclature 38
    K. A. Jensen, Chemical Laboratory, University of Copenhagen, Copenhagen, Denmark

Chemical Nomenclature in Britain Today 49
    R. S. Cahn and A. D. Mitchell, The Chemical Society, London, England

Chemical Nomenclature in the United States 55
    E. J. Crane, Chemical Abstracts, The Ohio State University, Columbus 10, Ohio

Basic Features of Nomenclature in Organic Chemistry 65
    Friedrich Richter, Beilstein-Institut, Frankfurt a. M., Hoechst, Germany

Organic Chemical Nomenclature, Past, Present, and Future 75
    P. E. Verkade, Laboratory of Organic Chemistry, Technical University, Delft, Netherlands

Work of Commission on Nomenclature of Biological Chemistry 83
    J. E. Courtois, Faculty of Pharmacy, Paris, France

Nomenclature in Industry 95
    H. S. Nutting, The Dow Chemical Co., Midland, Mich.

Development of Chemical Symbols and Their Relation to Nomenclature 99
    G. Malcolm Dyson, Loughborough, England

The Role of Terminology in Indexing, Classifying, and Coding 106
    James W. Perry, Massachusetts Institute of Technology, Cambridge, Mass.

INTRODUCTION

This symposium was a unique event. There have been conferences on chemical nomenclature, of which the outstanding one was the Congress of Geneva on organic nomenclature, held in 1892. But as far as my knowledge goes this series of papers presented in New York in 1951 constitutes the first symposium on chemical nomenclature held anywhere. The fortunate circumstance that the Diamond Jubilee of the AMERICAN CHEMICAL SOCIETY occurred immediately before the Sixteenth Conference of the International Union of Pure and Applied Chemistry made it possible to give the symposium a truly international character. Six different countries, Denmark, France, Germany, Great Britain, the Netherlands, and the United States, were represented among the eleven speakers. The three nomenclature commissions of the IUPAC were also represented, two by their presidents and one by its secretary. The letter of greeting from A. F. Holleman, former president of the Commission on the Nomenclature of Organic Chemistry, was written in his ninetieth year. W. P. Jorissen, former president of the Commission on the Nomenclature of Inorganic Chemistry, accepted an invitation to prepare a paper but was forced to withdraw for reasons of health. All the papers were read by their authors.

The symposium covered, in broad range and authoritative manner, the developments and problems of present-day chemical nomenclature. It is indeed gratifying that the papers can be published together as a number of the ADVANCES IN CHEMISTRY SERIES.

AUSTIN M. PATTERSON

Letter of Greeting

A. F. HOLLEMAN

University of Amsterdam, Amsterdam, Holland

I am very sorry not to be able to come to the United States and meet with you; my great age makes it impossible. So I must content myself with sending to you my cordial greetings and a few personal recollections from the time when I had the honor of being chairman of the Commission on the Reform of the Nomenclature of Organic Chemistry. I will add also some reflections on nomenclature.

You know that the International Union of Pure and Applied Chemistry decided in 1922 to nominate a commission for the revision of the Geneva nomenclature of 1892, the latter having become inadequate for the rapidly increasing number of organic compounds. The aim of the commission was (1) to revise the Geneva nomenclature, (2) to complete it for the multitude of new compounds, and (3) to respect as much as possible the existing nomenclature, especially as regards trivial names.

We had several meetings in Paris, and some changes in personnel were necessary in the course of years. Deliberations were somewhat difficult because of the language problem. At length, however, we had framed a number of rules and our project had to be submitted for the approval of the Union, which met in 1928 in The Hague. About fifty were present at this meeting, as I estimate. I explained the aim of the meeting, to discuss and possibly approve the rules that the commission had adopted after careful consideration. A well-known Englishman then announced that he had invented a new set of rules and asked to have them discussed. I answered that this was impossible, as they had not been studied beforehand by the commission. He answered: "Mr. Chairman, will you discuss my rules—yes or no?" On my negative answer he shut his portfolio with a bang and disappeared. The commission has never learned the rules that he wished to propose.

Then came a Frenchman. He said that he had not studied the rules that were proposed but that he wanted to present some ideas that had just come into his mind. I asked him to send his ideas to the commission, which would without doubt consider them thoroughly. We never received them. There were some other persons who uttered criticisms on our project, proving their incompetence in this matter. The end was that our project was neither approved nor rejected, so the whole discussion ended in nothing.

We were happier at the meeting of the Union in Liége in the year 1930. The meeting of the Union that had to approve our proposals agreed with them without discussion, and so we had reached our aim. As for myself, I thought that the time had come to retire, the more so as I had an eminent successor in the person of Verkade.

Of course I was perfectly aware that the task of the commission was not ended. On the contrary, it has to be permanent as long as organic chemistry is thriving. Its wonderful development in this century in its different branches necessitates a continual labor. Also, the rules were from the beginning far from complete and many series of compounds were not considered—e.g., the organosilicon compounds, extended rules for whose nomenclature have since been proposed.

Then there is the linguistic question. You of course want your rules to be generally used. It is long since Latin was understood by all scientists and it will certainly be several decades, perhaps even a century, before Esperanto is everywhere used by them. So, the nomenclature must be adapted to the several languages. I remember that Votoček proposed a whole other set of rules, adapted to the Czech language, and that in

1913 Istrati published a volume of no less than 1210 pages in the Romanian language about organic nomenclature in his country. It seems dubious that the spirit or the grammar of many nations will agree with your prefixes and suffixes. What can be thought of the Slavic languages, or Turkish, or Hebrew, or Japanese? So much work must still be done by the regional committees.

I end with sending to you my best wishes for the success of your efforts.

RECEIVED November 1951.

Some General Principles of Inorganic Chemical Nomenclature

HENRY BASSETT

Geological Survey Department, Dodoma, Tanganyika, Africa

The living and growing science of chemistry requires a nomenclature system that is expansible and mutable. Changes for the sake of conformity in different languages are made slowly and with difficulty. Empirical names are suitable for only the simplest substances and more descriptive nomenclature is needed. The expansion of knowledge of the nature of compounds, such as crystalline structures, leads to extension and perfection of the nomenclature system.

The ideal system of nomenclature should enable the maximum number of compounds to be named as unambiguously as possible. It should also be a friendly system, akin to that used in naming ordinary people rather than to that by which numbers are assigned to convicts. A purely mechanical system depending upon ciphers can never be satisfactory for general use, no matter how good it might be for indexing and similar purposes.

The fewer the number of individuals which have to be distinguished, the simpler the names which will suffice. Greater complexity becomes necessary with increasing numbers. This is general.

In the case of inorganic chemistry, increase in the number of known compounds has to a great extent run parallel with increased knowledge of the compounds. It is not surprising, therefore, that as extensions or modifications of earlier nomenclature became necessary, this has led to the dropping of trivial names which at one time were adequate and their replacement by others based upon some fundamental property of the compounds.

As chemistry is a live and growing science, no system of nomenclature can be expected to last indefinitely without need for alteration. The object must be to base the system of naming on principles which are least likely to require modification with increased knowledge. When such modification becomes necessary, it should not be shirked. The present fairly complex system has grown gradually from the simple system devised by Lavoisier and his colleagues. This system was able to deal very successfully with binary and ternary compounds and enabled simple differences in valency and states of oxidation to be distinguished. Since it was based upon the sound principle that binary compounds were formed by the union of positive (metal) and negative constituents it has been capable of continuous extension.

International Aspects of Nomenclature

As chemistry is international in character, it is desirable that its nomenclature should differ as little as possible in different languages, but it is not essential that it should be identical. It is almost inevitable that there will be some differences, especially where chemistry touches the ordinary lives of people. This applies to a number of the elements which happen to be important metals, known in some cases from prehistoric times. It is natural that the names of such metals differ in different languages, and it is probably useless and unwise to try to force the different languages to use the same international names for the metals in question. The symbols of these elements only affect chemists and the compromise made by the chemists of the early 19th century of using symbols for these metals derived from their Latin names has been thoroughly justified. As the symbols of the chemical elements represent the chemical alphabet from which chemical formulas are constructed, it is of great importance that the symbols used should be the same in all languages, even if there are some differences in the names of the elements. It is hoped that this uniformity of the symbols for the elements has been achieved at last.

The case of sodium and potassium has some resemblance to that of the ancient metals. After the isolation of these two metals from the alkalies which had long been known, their names in different languages were derived from different names of the alkalies, and so were obtained the alternative pair of names sodium and potassium and natrium and kalium (the symbols Na and K derived from the latter pair). Undue variations in the names of elements in different languages should be strongly opposed.

Many of the lighter elements have names which are unrelated to the symbols in certain languages, and more uniformity here is greatly to be desired. While any change in nomenclature causes trouble in connection with the indexing and abstracting of literature, a fundamental change of name of common chemical substances might cause trouble and expense not only to the indexing and abstracting organizations, but also to chemical and even other industries. These effects might outweigh the advantages to be gained from greater uniformity of nomenclature, especially in the larger and more industrialized countries.

All living languages tend to change slowly, but every effort should be made to prevent unauthorized changes being made in chemical names—especially as such changes would almost certainly differ in different languages.

There is no question that for convenience and ease in reading of chemical papers in foreign languages it is most desirable that the nomenclature should be as similar as possible.

In the Romance languages, the more electronegative constituent is cited first when naming salts and other compounds and even written first in the formulas (*chlorure de sodium*, ClNa). It is unfortunate that the idea has grown up that this is inherent in the construction of the language. It is probably little more than custom which could be changed without much difficulty. After all, in English the old system of naming salts was very much the same and people spoke of chloride of silver, sulfate of lead, etc.

Limitations of Empirical Names

The very least that can be expected of the name of a chemical compound is that it should enable a chemist to write down its empirical formula. If the latter is known, the compound can always be named by citing the number of each kind of atom present. A name of this sort is sufficient in many cases, but usually only for the simplest compounds. Since the names will be used for indexing purposes, it is important that the elements should always be cited in the same order and, though based on the principle of putting the most electropositive element first and the most electronegative last in English and related languages, the order has to be an arbitrary one owing to the large number of possible elements which are involved. As far as is possible, a rule of nomenclature should depend on some general principle but should be laid down in such a way as to make its interpretation definite.

Compounds may become of technical importance before their precise composition in the pure state is known. Such substances may be given names, which will usually have to be trivial ones.

In other cases, compounds may be given trivial names because the correct systematic ones would be too cumbersome. This often happens when compounds have important practical applications. It is probably not possible to prevent this practice completely, but it should not be encouraged. When definite compounds are given trivial names, a glossary becomes necessary to interpret these names. During the past 150 years, chemistry has had the great advantage, as compared with the more descriptive sciences, that its system of nomenclature has, on the whole, provided its own interpretation. This advantage would be lost with any widespread adoption of trivial names.

Names based solely upon the empirical formulas are inadequate when more than one compound is known with the same empirical formula. This difficulty can be overcome where the substances in question have different molecular formulas. But many isomeric compounds of the same molecular weight gradually became known and it was also clear that names based simply upon empirical or molecular formulas were not very helpful. It was thus appreciated from the beginning of the modern chemical period and of modern chemical nomenclature that a satisfactory name for a chemical compound must tell something about its structure as well as about its composition.

Methods for indicating the presence of definite groups of atoms common to a number of compounds such as "sulfate" and "carbonate" and of differing valency states of the same metal such as "ferrous" and "ferric" go back to Lavoisier.

Need for Descriptive Names

All chemists are agreed that chemical names must tell something about the nature of the compounds, but there is considerable difference of opinion as to just how much a name should tell. The answer appears to be that the name should be as simple as the circumstances justify, and therefore that the minimum amount of information should be given about the nature and structure of the compound. It follows that different names, giving varying amounts of information, may be needed for the same compound. This has become especially true of recent years, now that so much is known about the structure of compounds in the crystalline state. Today it is often a question not merely of giving one sound name to a compound but several, all equally sound within the limits of the information they give.

Consider the case of cupric chloride. Under many circumstances of a general nature, such as reactions using solutions of the compound, the name cupric chloride or, better still, copper dichloride suffices. If referring especially to the hydrated solid salt, either cupric chloride dihydrate or copper dichloride dihydrate is used. The full structure of the crystalline hydrated chloride is known, and if matters involving this structure were being dealt with, then the full name should be given as *trans*-copper dichloride dihydrate since actual flat molecules
$$\begin{matrix} H_2O & Cl \\ & Cu & \\ Cl & OH_2 \end{matrix}$$
are present in the crystal.

Views about the nature of compounds change as more is learned about them. One reason for keeping names as simple as possible is that they are then less likely to need alteration if increased knowledge shows that previous ideas were not quite correct.

Complications of Crystalline State

One of the major problems of inorganic nomenclature concerns the many compounds which exist only in the crystalline state. Solution, fusion, or volatilization cause profound changes which are tantamount to decomposition of such compounds.

The atomic arrangement in the crystalline solid is known in a great many cases, but in a very large proportion of these it would be extremely difficult to give a fairly short and simple name which would describe this arrangement.

The difficulty of the problem is enhanced because it is now known that the bonds between atoms in compounds and in crystals can be partly ionic and partly covalent in character, and not simply one or the other. This applies in particular to many mixed oxides so that aluminates, titanates, zincates, etc., can be regarded as built up entirely from metal and oxygen ions, though to some extent negative aluminate, titanate, etc., ions seem to be present.

Two different names may be equally correct for the same class of compound according to circumstances, although this may really be connected with a change of structure brought about, for instance, by hydration. Thus the anhydrous aluminates, titanates, stannates, zincates, etc., have more the character of ionic double oxides. Soluble members of these groups, which in practice means only the alkali metal compounds, when hydrated or in solution are true salts with complex hydrated anions—compounds such as $NaAl(OH)_4$ and $K_2Sn(OH)_6$. It would seem better to name two such different types of compounds in different ways, the former as double oxides and the latter as complex compounds, although, for purposes of indexing, the older names have a certain value.

Naming New Compounds

It is easier to give good names to compounds which have long been known than to those which have only been recently discovered, especially when the latter belong to entirely new classes of compounds. Responsibility belongs to research workers in this connection. Caution should be used in naming new compounds, especially when they are of entirely new types.

The existing rules of nomenclature will usually permit satisfactory temporary names based upon empirical or molecular formulas.

More distinctive names should not be given until the reactions of the new compounds and their relationship to other compounds are sufficiently well known.

RECEIVED August 1, 1951.

Nomenclature of Coordination Compounds and Its Relation to General Inorganic Nomenclature

W. CONARD FERNELIUS

The Pennsylvania State College, State College, Pa.

> The pattern of chemical combination represented by coordination compounds is one of common occurrence. The adoption of a basic plan of nomenclature for this fundamental pattern of chemical combination is desirable. The basic philosophy underlying the various schemes for the naming of coordination compounds is reviewed briefly. The causes of apparent contradictions are discussed. The practices for the nomenclature of coordination compounds suggested by the Commission on the Reform of Inorganic Chemical Nomenclature of the International Union of Chemistry (1940) previously were formulated into a set of rules (1948). These rules are now modified in the light of criticism and further study and are extended to cover situations not previously included. Examples are given to show how the practices prescribed by the rules may be extended.

From the standpoint of nomenclature, inorganic chemical compounds may be divided into the following major divisions: binary, ternary, and higher order. The development to date of nomenclature systems for each of these classes is summarized in the following statements:

Binary Compounds (NH_3, KCl, etc.). The situation is generally satisfactory except for intermetallic compounds and, possibly, minerals.

Ternary Compounds (H_3PO_4, $Na_2S_2O_3$, etc.). The recommendations of the Commission on the Reform of Inorganic Chemical Nomenclature (20) of the International Union of Pure and Applied Chemistry are reasonably comprehensive and have been followed widely. While the various prefixes and terminations do not mean the same thing for all elements, they do have generality within a given family of the periodic table. Those cases where the inconsistencies are most evident are of long standing and are generally familiar to all chemists.

Higher Order Compounds. The development of nomenclature for these compounds has lagged far behind that for the other classes and there are many urgent problems. A detailed consideration of the situation is in order.

Among compounds of higher order, the following broad types are recognized:

1. Solvates: hydrates, $CaCl_2 \cdot 6H_2O$; ammoniates, $AgCl \cdot 2NH_3$; etc.
2. Molecular addition compounds: $(CH_3)_3N \cdot BF_3$
3. Double salts: $3NaF \cdot AlF_3$
4. Iso- and heteropoly acids and their salts: $H_6W_6O_{21}$, $K_3P(Mo_3O_{10})_4$
5. Coordination compounds

These types are not all sharply divided, but many compounds may be considered to belong to two or more of these classifications. Although schemes of nomenclature are available for all of these types, the schemes are reasonably satisfactory for only three (solvates, double salts, and iso- and heteropoly acids and salts). This situation is the more regrettable because so large a proportion of the total number of known compounds belongs to the class of higher order compounds and modern research is rapidly enlarging both the number and complexity.

Most students of chemical nomenclature will agree that a broad general pattern which is capable of extension to diverse types of compounds is preferable to a number of specific patterns of limited extension. Inasmuch as most compounds of higher order can and should be looked upon as coordination compounds, coordination compounds encompass an extremely broad field. The development of a satisfactory scheme of nomenclature for these compounds may solve many nomenclature problems. If coordination compounds represent a fundamental pattern of chemical combination, there should be a sound basic plan for the nomenclature of this broad class of compounds.

Theory of Werner

Alfred Werner (49) was the first to systematize coordination compounds in a satisfactory manner. He distinguished two types of valences which he termed main (or principal) and auxiliary valence. The former represented those manifestations of chemical affinity which result in the formation of simple binary compounds: H_2O, $CuCl_2$, NH_3, $CrCl_3$, etc. The latter represented those manifestations of (residual) chemical affinity which are able to bring about the stable union of molecules as if the molecules were themselves radicals able to exist as independent molecules: $CuCl_2 \cdot 4H_2O$, $CrCl_3 \cdot 6NH_3$, etc.

Werner considered that when the binding capacity of an elementary atom appears exhausted, it can still link up with, or coordinate, other molecules and build up more complex structures, but that there is an upper limit to this process. The maximum number of atoms, radicals, or molecular groups—independent of their charge—which can be directly linked with a central atom he called the **coordination number** (C.N.) of that atom: C.N. 4, $[Pt(NH_3)_4]Cl_2$, $K_2[PtCl_4]$; C.N. 6, $[Pt(NH_3)_6]Cl_4$, $K_2[PtCl_6]$; C.N. 8, $K_4[Mo(CN)_8]$. The central atom with its surrounding coordinated atoms or groups (coordination sphere) constitutes, according to Werner, a unit which is not a salt, but it may be a radical which can combine with other radicals to form a salt. The effective charge of the coordinated group depends both in magnitude and sign upon the nature of the atoms or groups attached to the central atom: $[Co(NH_3)_6]Cl_3$, $[Co(NH_3)_5Cl]Cl_2$, $[Co(NH_3)_4(NO_2)_2]NO_2$, etc. Finally, the coordinated atoms or groups are arranged symmetrically in space: C.N. 4, either tetrahedral or planar; C.N. 6, octahedral; C.N. 8, as yet not thoroughly studied.

The proposals of Werner (49) about the structure of coordination compounds were amply justified by the chemical reactions exhibited by them, by the identification of geometrical isomers, and by the resolution of certain compounds into their optical antipodes. Modern methods for the determination of structures and for studying the nature of charged or neutral species in solution have demonstrated the fundamental soundness of Werner's views.

Recently, increased activity in the study of the fundamental nature of coordination compounds and in the use of their special properties in many technical applications indicates an urgency about reaching agreement on general principles concerning their nomenclature. In coordination compounds there are many combinations of organic groups or molecules with inorganic units. Hence, the solution of problems of nomenclature may aid greatly in keeping workers in the two divisions of chemistry informed of the other's problems and of the need for reaching nonconflicting solutions.

Terms Descriptive of Coordination Compounds

Before discussing the problems of the nomenclature of individual coordination compounds, it is desirable to review the terms which have been used to define either the position of the structural unit or its function in the coordination compound. Every coordination compound contains one or more complex units consisting of a central atom (the center of coordination, actually or potentially a cation) surrounded by a number of other atoms (coordinated groups). These other atoms may bring with them a charge (almost always negative) and/or additional atoms which comprised an ion or neutral molecule prior to the coordination. Thus, the complex or coordination unit considered as a whole may be a cation, an anion, or a neutral unit. A coordination unit, without regard to its charge, is here referred to as a coordination **entity** (often called a **complex**). Many authors have used the term **complex compound** in preference to **coordination compound**. However, the use of a general term in so specific a sense is unwise. Many an author has described a compound as complex (meaning complicated) only to have others assume he was implying that it was a coordination compound. This confusion is unfortunate and should not be allowed to continue.

Centers of coordination differ in the number of bonds which they form with coordinated groups. The term **coordination number** refers to the number of groups bonded to the center of coordination.

Sidgwick (42) has discussed the significance of the maximum coordination number which any atom can exhibit and Bjerrum (1) has distinguished between the characteristic and the actual coordination number.

Coordination numbers of 4 and 6 are most common. Those of 2 and 8 occur frequently, while those of 3, 5, and 7 are comparatively rare. The crystallographer uses the term coordination number in a somewhat different sense. The term is used by the crystallographer to designate the number of nearest neighbors surrounding any atom or ion. For crystals containing ions such as $[Ni(NH_3)_6]^{++}$ and $[PtCl_6]^{--}$, the crystallographer and the chemist use the term coordination number as applied to nickel and platinum synonymously. For crystals containing networks of silicon atoms (ions) surrounded by oxygen atoms (ions) in silica and in the various silicates, or for platinum atoms surrounded by sulfur atoms in platinum sulfide, the two uses of the term coincide. For simple binary compounds having ionic lattices such as lithium fluoride, the crystallographer uses the term in a sense completely different from that of the chemist.

Further, it should be noted that, in contrast to the crystallographer, the chemist frequently is interested in the existence of a coordination entity in the liquid or dissolved state. In some instances there is reason to feel that the coordination number, or at least the configuration, is not the same in the solid as in the dissolved state.

The coordinated ions or molecules have been referred to as ligands or ligates, addenda, and adducts. Their electron pair donor character has been stressed since the recognition of the electronic nature of chemical combination. Further, a given ligand like ethylenediamine ($H_2NCH_2CH_2NH_2$) or oxalate ion ($^-OOCCOO^-$) may attach itself to a center of coordination at more than one place (donor atom) on the ligand group. Thus, on the basis of the number of points of attachment, a ligand may be described as follows:

Unidentate	Single point of attachment	NH_3, F^-, CN^-
Bidentate	Two points of attachment	$H_2NCH_2CH_2NH_2$, $^-OOCCOO^-$
Terdentate	Three points of attachment	$HN(CH_2CH_2NH_2)_2$, $HN(CH_2COO)_2^{--}$
Quadridentate	Four points of attachment	$N(CH_2CH_2NH_2)_3$, $(—CH_2NRCH_2COO)_2^{--}$
Quinquadentate	Five points of attachment	$^-OOCCH_2NRCH_2CH_2N(CH_2COO)_2^{--}$
Sexadentate	Six points of attachment	$\left(—CH_2SCH_2CH_2N=CH\underset{}{\bigcirc}^O\right)_2^{--}$

The adjective **chelate** (24) was proposed to designate the bidentate character of a group but has been generalized to include all polydentate groups and has been used loosely as a noun both for the chelate group and for the entity resulting from the combination of chelate groups with a center of coordination.

Following the terminology introduced by Ley (21), the coordination of a bidentate group which has both a charged and an uncharged position for coordination, produces an **inner complex compound** (or **internally complex compound**):

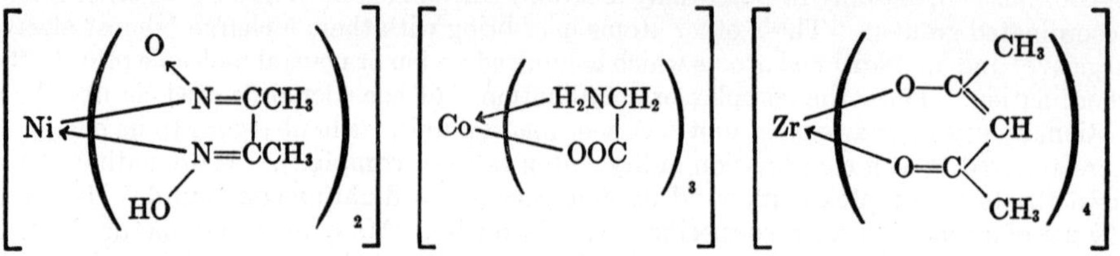

$$\left[Ni \left(\begin{array}{c} O \\ N=CCH_3 \\ N=CCH_3 \\ HO \end{array} \right) \right]_2 \quad \left[Co \left(\begin{array}{c} H_2NCH_2 \\ | \\ OOC \end{array} \right) \right]_3 \quad \left[Zr \left(\begin{array}{c} CH_3 \\ O-C \\ CH \\ O=C \\ CH_3 \end{array} \right) \right]_4$$

In view of the general objection to the term complex compound, it would be desirable to have an alternate expression for inner complex compound. For the purposes of this paper the expression **inner coordination compound** will be used.

Care should be taken to distinguish these terms from the expressions "inner orbital" and "outer orbital" complex ions which have been introduced recently for an entirely different purpose (45).

Although the term inner coordination compound is usually restricted to the meaning given above, some writers have used it as synonymous with **nonionic coordination compound** to include nonchelate compounds like $[Co(NH_3)_3(NO_2)_3]$ and partially chelated compounds:

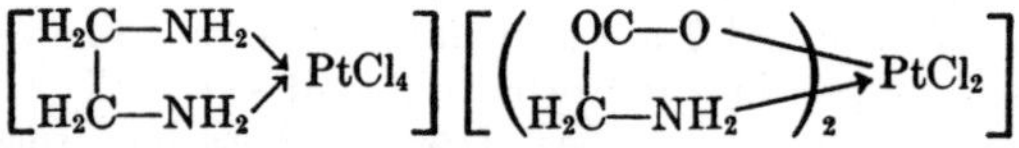

$$\left[\begin{array}{c} H_2C-NH_2 \\ | \\ H_2C-NH_2 \end{array} PtCl_4 \right] \quad \left[\left(\begin{array}{c} OC-O \\ H_2C-NH_2 \end{array} \right)_2 PtCl_2 \right]$$

The value of these extensions is very doubtful. A few authors (5) recognize two types of inner coordination compounds. An inner coordination compound of the **first order** is a substance like the examples given which is a neutral body without saltlike properties. The cations forming them have coordination numbers equal to twice the charge on the ion. An inner coordination compound of the **second order** is a substance arising from the reaction of singly charged bidentate groups with a simple cation but in which the coordination entity bears a charge:

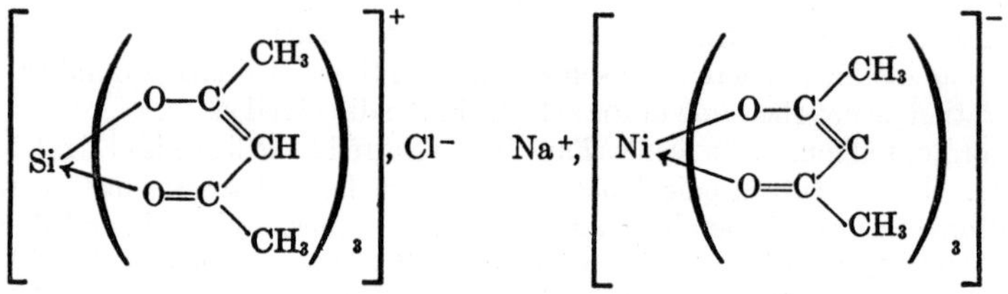

$$\left[Si \left(\begin{array}{c} CH_3 \\ O-C \\ CH \\ O=C \\ CH_3 \end{array} \right)_3 \right]^+, Cl^- \quad Na^+, \left[Ni \left(\begin{array}{c} CH_3 \\ O-C \\ C \\ O=C \\ CH_3 \end{array} \right)_3 \right]^-$$

For the cations forming this latter type of compound the coordination number is either less or greater than twice the charge on the ion.

Quadridentate coordinating groups may also lead to neutral bodies to which the term inner coordination compound could well be applied:

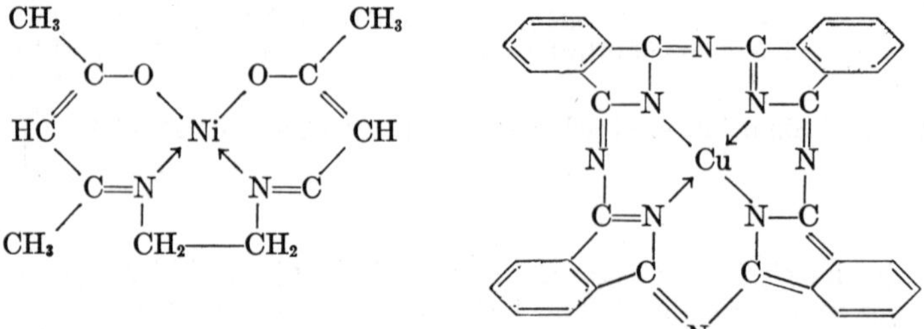

Pfeiffer (*31, 33*) proposed a classification of chelate rings based on the kind (and number) of bonds involved in linkages to the metal:

Principal Valence Ring	Auxiliary Valence Rings		
	First Kind	Second Kind	Third Kind
M⟨O−C=O / O−C=O	M⟨O−C=O / N−CH₂ / H₂	M⟨H₂N−CH₂ / N−CH₂ / H₂	⟩M⟨H,O / O,H⟩M⟨

(For the significance of the terms principal and auxiliary see the next paragraph.)

Methods for Writing Formulas of Coordination Compounds

The ease of naming coordination compounds is closely related to the effectiveness of the method of writing the proved (or likely) formulas for such compounds. Within the coordination sphere, Werner regarded the bonds between the coordination center and the coordinated groups as arising from principal valence forces (*Hauptvalenzkräfte*) or auxiliary valence forces (*Nebenvalenzkräfte*). The principal forces which reduced the charge on the central atom were represented by solid lines and the auxiliary forces, which had no effect on the charge, by dotted lines:

$$\begin{bmatrix} H_3N & NH_3 \\ H_3N\text{---}Pt\text{---}NH_3 \\ H_3N & NH_3 \end{bmatrix} Cl_4 \quad \begin{bmatrix} H_3N & Cl \\ H_3N\text{---}Pt\text{---}Cl \\ H_3N & NH_3 \end{bmatrix} Cl_2$$

$$\begin{bmatrix} H_3N & Cl \\ H_3N\text{---}Pt\text{---}Cl \\ Cl & Cl \end{bmatrix} \quad \begin{bmatrix} H_2CNH_2 & OOC \\ & Pt & \\ COO & NH_2CH_2 \end{bmatrix}$$

$$\begin{array}{c} C_2H_5 \quad Br \quad C_2H_5 \\ \diagdown \diagup \diagdown \diagup \\ Au \quad Au \\ \diagup \diagdown \diagup \diagdown \\ C_2H_5 \quad Br \quad C_2H_5 \end{array}$$

This representation has many commendatory aspects and continues in use today, especially among the Germans. The system breaks down when it is desired to represent the formulas of compounds where the number of negatively charged groups linked to the central ion exceeds the charge on the central ion $K_2(PtCl_6)$. It may be considered that molecules of the salt potassium chloride are linked by auxiliary valence forces:

$$\begin{array}{c} KCl \quad Cl \\ \diagdown \diagup \\ KCl\text{---}Pt\text{---}Cl \\ \diagup \diagdown \\ Cl \quad Cl \end{array}$$

The distinction between principal and auxiliary forces may be forgotten and a dot (·) used:

$$K_2 \begin{bmatrix} Cl\cdot & \cdot Cl \\ Cl\cdot Pt\cdot Cl \\ Cl\cdot & \cdot Cl \end{bmatrix}$$

or a solid line used to represent a bond of either type between atoms:

$$\begin{array}{c} Cl \quad NH_3 \\ \diagdown \diagup \\ Pt \\ \diagup \diagdown \\ H_3N \quad Cl \end{array}$$

Electron Concept. With the rise of the electron concept of valence, it was recognized that a covalent bond might arise in such a way as to neutralize the charges of two oppositely charged ions (normal covalence) or to create a partial charge on a hitherto uncharged atom (coordinate covalence). The former was represented by a solid line, the latter by an arrow with the head indicating the direction of the electron transfer ($B \rightarrow A = B \overset{+\;-}{\text{—}} A$):

$$\begin{bmatrix} H_3N & & Cl \\ & \searrow & \\ H_3N \rightarrow & Pt & — Cl \\ & \nearrow & \\ H_3N & & Cl \end{bmatrix} Cl \qquad \begin{bmatrix} H_2CNH_2 & & H_2NCH_2 \\ & \searrow \;\swarrow & \\ & Pt & \\ & \nearrow \;\searrow & \\ COO & & OOC \end{bmatrix}$$

$$\begin{array}{c} C_2H_5 \quad\;\; Br \quad\;\; C_2H_5 \\ \diagdown \;\diagup \;\diagdown \;\diagup \\ Au \quad\;\; Au \\ \diagup \;\diagdown \;\diagup \;\diagdown \\ C_2H_5 \quad\;\; Br \quad\;\; C_2H_5 \end{array}$$

While some authors (*29*) object to this distinction between normal and coordinate covalent bonds, others (*34, 36, 46, 48*) insist that it has value, if for no other reason than the ready computation of the charge on the coordination entity.

Modifications. From time to time there have been modifications of this method to meet certain needs. Some of these deserve comment. For example:

$$\left[M \!\!\underset{}{\overset{}{\Big\langle}}\!\! \begin{pmatrix} -NH_2CH_2 \\ | \\ -OOC \end{pmatrix} _3 \right]$$

This formula indicates that in taking three glycinate ions per metal ion, three negative charges and six coordinating positions have also been supplied. An alternate and frequently used method of representing the same compound is:

$$\underset{3}{M} \!\!\diagup\!\! \begin{matrix} NH_2CH_2 \\ | \\ OOC \end{matrix}$$

The $\dfrac{M}{3}$ indicates $1/3$ of an atom of M or that three chelate groups are attached to atom M.

The formula:

$$\begin{bmatrix} Co & \overset{(NH_3)_5}{\diagdown} \\ & \diagdown \\ & O\text{———}SO_2\text{—}O^- \end{bmatrix}^{++}$$

indicates that only one of the charges on the $SO_4{}^{--}$ is neutralized within the coordination entity and that the other must be neutralized externally. This is an artificial distinction and usually is not made. The formula $[Co(NH_3)_5SO_4]^+$ is adequate to represent the structure and properties of the coordination ion if the justifiable assumption of 6-coordinate cobalt(III) is made.

Perhaps the most complicated formulations were those used by Morgan and Smith (*25*) in representing the formulas of certain coordination compounds containing dye molecules. The normal salt composed of the hexamminecobalt(III) cation and three 1-nitroso-2-naphtholate ions is represented thus:

On heating, this is converted into tris(1-nitroso-2-naphtholato)cobalt(III) which is represented:

$$\left[\left\{\begin{array}{c}\text{NO}\\ \parallel\\ \text{=O}\end{array}\right\}\text{Co}\right]$$

The symbolism indicates that each 1-nitroso-2-naphtholate ion is attached to the central cobalt by one principal and one auxiliary valence bond and that the total of three such naphtholate units supply bonds for the six coordinate positions of the central cobalt.

A more complicated situation is represented by the formula:

$$\left[\left\{\begin{array}{c}\text{NO}\\ \parallel\\ \text{=O}\\ \text{COO}\end{array}\right\}\text{Co}\right]_3 [\text{Co(NH}_3)_5]$$

The =NO⁻ and =O groups stand in the same relationship to the central cobalt ion as in the previous formula. Of the three bonds from the COO⁻ groups, one is coordinated to the cobalt along with five ammonia molecules, while the other two neutralize the charge remaining associated with this unit. A situation as complicated as this is seldom encountered.

Development of Nomenclature Systems

A brief review of the historical development of the nomenclature of coordination compounds will indicate the basis for the different proposals and point out the origin of conflicts in practice. Prior to the proposal of any general system, individual coordination compounds were known by specific trivial names, often based on color or the name of the discoverer: Magnus's green salt, Roussin's red and black salts, violeo series, roseo salts, etc. Names based on color are still in common use. However, even the value gained by basing a name on such a definite property as color has been lost by overextension. For example, because praseo series was used to designate the green compounds trans-[Co(NH$_3$)$_4$X$_2$]X and trans-[Co(en)$_2$X$_2$]X, the orange compounds trans-[Ru(NH$_3$)$_4$-X$_2$]X also are spoken of as belonging to the praseo series (13).

Two attempts at generality in naming were made independent of the proposals of Werner. The first of these arose from a consideration of the coordination compound as a kind of double salt with the -o- and -i- indicating a lower and a higher state of oxidation as the familiar -ous and -ic terminations.

Na$_4$[Co(NO$_2$)$_6$]   Sodium cobaltonitrite   K$_4$[Mo(CN)$_8$]   Potassium molybdocyanide
Na$_3$[Co(NO$_2$)$_6$]   Sodium cobaltinitrite   K$_3$[Mo(CN)$_8$]   Potassium molybdicyanide

This system is capable of only limited extension.

The second proposal originates from the idea that corresponding to every oxygen salt there are sulfur, chlorine, etc., salts which are thought of as products of the substitution of oxygen by sulfur, chlorine, etc., or as products of comparable origin (10–12, 37).

K$_2$O + B$_2$O$_3$ → K$_2$O·B$_2$O$_3$ or 2KBO$_2$ Potassium borate
K$_2$S + B$_2$S$_3$ → K$_2$S·B$_2$S$_3$ or 2KBS$_2$ Potassium **thio**borate
KF + BF$_3$ → KF·BF$_3$ or KBF$_4$ Potassium **fluoro**borate
K$_3$N + BN → K$_3$N·BN or K$_3$BN$_2$ Potassium **ammono**borate

The names given these compounds reflect this supposed relationship. This system has as much generality as the widely accepted schemes for naming salts of oxygen acids.

Note. The prefix **ammono** has been used in the names of a number of nitrogen compounds:

H_4PN_3	Ammonophosphoric acid
$CaCN_2$	Calcium ammonocarbonate
$K_2[Zn(NH_2)_4]$	Potassium ammonozincate
$HgCl(NH_2)$	Ammonobasic mercuric chloride

However, it is obviously a poor choice because it throws the emphasis toward ammonia (the solvent in which these compounds were prepared) rather than nitrogen. Unfortunately, all forms based on the root nitrogen already had acquired specific and completely different significance.

Proposals of Werner. The system of nomenclature proposed by Werner (50) did not suffer the limitations of these schemes, since it was capable of wide extension. To name any coordination entity, Werner listed the ligands (negative first with termination -o followed by the neutral ones without characteristic termination) and then the center of coordination. To the root characteristic of the central atom was added a termination characteristic of the oxidation state of this atom:

(I, **a**; II, **o**; III, **i**; IV, **e**; V, **an**; VI, **on**; VII, **in**; VIII, **en**)

and finally, if the coordination entity was an anion, the termination **-ate**.

$[Cr(NH_3)_6]^{3+}, 3NO_3^-$	Hexamminechromi nitrate
$[Pt(H_2O)_2(NH_3)_4]^{4+}, 4Cl^-$	Diaquotetrammineplate chloride
$[Co(NO_2)Cl(en)_2]^+, Br^-$	Nitrochlorobis(ethylenediamine)cobalti bromide
$H^+, [AgCl_2]^-$	Hydrogen dichloroargentaate
$3K^+, [Fe(CN)_5(H_2O)]^{---}$	Potassium pentacyanoaquoferroate
$2Na^+, [SSO_3]^{--}$	Sodium thiotrioxosulfuronate

Similar but different names will be obtained depending upon whether they are derived on the basis of a system of compounds or on the pattern of Werner:

K_2PtCl_4	Potassium chloroplatinite	Potassium tetrachloroplat**oate**
K_2PtCl_6	Potassium chloroplatinate	Potassium hexachloroplat**eate**

Werner introduced some other practices such as the use of μ to signify a bridging group, of **-ol** for bridging hydroxyl groups, and of *cis* and *trans* or numbers to designate isomers.

It is difficult to understand why the Werner nomenclature was not accepted more widely. It had most of the aspects of a good system. Perhaps the difficulty of remembering eight different endings was too great. Perhaps the indifference of inorganic chemists regarding nomenclature problems until recent years is responsible. Another factor may have been the lack of workers in the field of coordination compounds. At any rate, writers on coordination compounds fell into the practice of contenting themselves with formulas and omitting names.

The practice is somewhat varied as to the use of the **-o** termination. Few authors have made clear, straightforward statements of the rules they were following in the naming of coordination compounds. Hence, the rule has to be deduced from the observed practice. However, Werner (49), Weinland (47), Pfeiffer (32), Gmelin (14), Schwarz (41), and Wittig (53) have been uniform in their use of **-o** with negative groups and in its omission from neutral groups. The German (original) edition of Hückel's book on inorganic structural chemistry (16) follows this practice. However, the nomenclature practices of the English edition (17) are inconsistent and contradictory; further, they involve hybridizations found nowhere else. Sutherland (44) and Jaeger (19) consistently used a terminal **-o** for both negative and neutral groups.

Mellor (22) uses ammino and amino (for RNH_2) (also aquo), but not consistently. Otherwise he does not give neutral groups an **-o** ending. Ephraim (6) states "... groups whether negative radicals or neutral groups, are indicated by the addition of the ending **-o** to their ordinary names." However, subsequently the terminal **-o** is used or omitted with neutral groups indiscriminately. Hein (15), who has written the most recent compilation in this field, states that all ligands other than ammine have the ending **-o**. However, throughout the book, no neutral group other than aquo is given a terminal **-o**.

In naming coordination compounds, for the most part, there has been virtually no effort to incorporate systematic names for the organic groups.

As an example, consider the wide variance in the names given to the metal derivatives of the condensation products of ethylenediamine with 2,4-pentanedione (acetylacetone) and salicylaldehyde.

This has led to the nearly consistent use of many archaic and trivial names for organic groups. This practice should be discontinued except where the complications resulting would be too great for convenience.

Later Modifications. The Stock proposal (*26, 38, 43*) that Roman numerals in parentheses (instead of -ous and -ic terminations) be used to designate oxidation state found immediate acceptance. Its extension to coordination has been received with enthusiasm. These names are being used extensively today in harmony with the recommendations of the IUPAC commission report (*20*).

Perhaps the most important aspect of the IUPAC report was the stimulation of interest in the problems of nomenclature, although the report left many problems with no solution. Solutions to some of these problems may be suggested on the basis of practices originating with Werner. Solutions to others require independent study. Two groups of workers have studied the general problem of the nomenclature of coordination compounds: Fernelius, Larsen, Marchi and Rollinson (*8*), and Ewens and Bassett (*7*). The suggestions embodied in the publications of these workers are incorporated in the rules given below.

Proposed Rules and Comments

Rule 1. Order of listing ions. The cation is named first, followed by the anion.

Examples:

$[Co(NH_3)_6]^{+++}$, $3Cl^-$	Hexamminecobalt(III) chloride
$4K^+$, $[Fe(CN)_6]^{----}$	Potassium hexacyanoferrate(II)
$[Pt(NH_3)_4]^{++}$, $[Pt(CN)_4]^{--}$	Tetrammineplatinum(II) tetracyanoplatinate(II)

COMMENT. This is simply an extension of the practice followed in naming simple salts.

Rule 2. Characteristic endings of coordinated groups. The endings of the names of coordinated groups are: positive, **-ium**; neutral, none; and negative, **-o**.

Examples:

$NH_2NH_3^+$	Hydrazinium
H_2O	Aqua (instead of aquo)
NH_3	Ammine
CN^-	Cyano
CO_3^{--}	Carbonato

COMMENT. There are only a few instances where a positively charged group is known to coordinate to a central ion. Most of these involve polydentate groups with a free amino group carrying a proton. These would normally be named as ammonium compounds:

$$\begin{bmatrix} Cl & NH_2CH_2 \\ & \diagdown \diagup \\ & Pt \\ & \diagup \diagdown \\ Cl & NH_2CH \\ & | \\ & CH_2NH_3 \end{bmatrix}^+, \; Cl^-$$ (2,3-Diaminopropylammonium)- dichloroplatinum(II) chloride

It seems logical to extend the **-ium** termination to the few instances not included in the ammonium group: $(M)H_2NNH_3^+$, $(M)NO^+$ (nitrosylium).

Considering the fact that one is always alert either in the spoken or written name to the charged or uncharged character of coordinated groups, it seems desirable to limit

Table I. List of Coordinated Groups

A. **Positively Charged (Cationic) Groups**
  1. Inorganic
      Hydrazinium, $H_2NNH_3^+$
      Nitrosylium, NO^+
      Tetramminedihydroxocobaltium(III), $(NH_3)_4Co(OH)_2^+$

  2. Organic
      2-Aminoethylammonium, $H_2NCH_2CH_2NH_3^+$
      2-(2-Aminoethylthio)ethylammonium, $H_2NCH_2CH_2SCH_2CH_2NH_3^+$
      2,3-Diaminopropylammonium, $H_2NCH_2CHNH_2CH_2NH_3^+$

B. **Neutral Groups**
  1. Inorganic
      Ammine, NH_3, am
      Antimony trichloride, $SbCl_3$
      Aqua, H_2O
      Carbonyl, CO
      Hydrazine, H_2NNH_2
      Hydroxylamine, H_2NOH
      Nitrosyl, NO
      Phosphorus oxychloride, $POCl_3$
      Phosphorus sulfochloride, $PSCl_3$
      Phosphorus trichloride (trichlorophosphine), PCl_3
      Phosphorus trifluoride (trifluorophosphine), PF_3

  2. Organic
      Acetonitrile, CH_3CN
      Alcohols, see methanol, ethyleneglycol, etc.
      Amines, see methylamine, ethylenediamine, etc.
      Aminoalcohols, see ethanolamine, diethanolamine, etc.
      (2-Aminoethyl)methylsulfide, $CH_3SCH_2CH_2NH_2$
      2-Aminopyridine
      Antipyrine
      Arsines, see triethylarsine
      2,2'-Bipyridine (α,α'-dipyridyl) (dipy)
      Bis(2-aminoethyl)sulfide, $S(CH_2CH_2NH_2)_2$
      2,2'(or 4,4')-Bithiazole,

      2,3-Butanediamine, $CH_3CH(NH_2)CH(NH_2)CH_3$, bn
      Catechol
      trans-1,2-Cyclopentanediamine, (cptdin)
      Diethanolamine, $(HOCH_2CH_2)_2NH$
      Diethylenetriamine, $HN(CH_2CH_2NH_2)_2$, dien
      Diethylether, $(CH_3CH_2)_2O$
      Diethylsulfide, $(CH_3CH_2)_2S$
      2,6-Di-2'-pyridylpyridine (α,α',α''-terpyridyl) (trpy)
      Ethanol, C_2H_5OH
      Ethanolamine, $HOCH_2CH_2NH_2$
      Ethers, see diethylether, etc.
      Ethylamine, $C_2H_5NH_2$
      Ethylene, $H_2C=CH_2$
      Ethylenediamine, $H_2NCH_2CH_2NH_2$, en (ene)
      Ethyleneglycol, $HOCH_2CH_2OH$
      Ethylenethiourea, see 2-imidazolidinethione
      2-Imidazolidinethione, 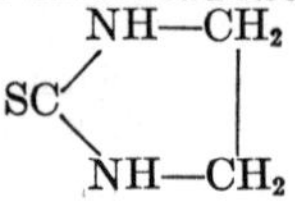 (etu)
      Methanol, CH_3OH
      Methylamine, CH_3NH_2
      Methylisocyanide, CH_3NC
      2,3-Pentanediamine, $CH_3CH(NH_2)CH(NH_2)CH_2CH_3$
      1,10-Phenanthroline (*o*-phenanthroline)
      o-Phenylenediamine

Table I. List of Coordinated Groups (continued)

B. **Neutral Groups** (continued)

2. Organic (continued)
   Phosphines, see trimethylphosphine
   1,2-Propanediamine (propylenediamine), $CH_3CH(NH_2)CH_2NH_2$, pn (propin)
   1,3-Propanediamine (trimethylenediamine), $H_2N(CH_2)_3NH_2$, (tme)
   1,2,3-Propanetriamine, $H_2NCH(CH_2NH_2)_2$, ptn, (tn), (triprop)
   Propylenediamine, see 1,2-propanediamine
   Pyrazine
   Pyridine, C_5H_5N, py
   Quinoline
   Stibines, see triethylstibine
   Sulfides, see diethylsulfide
   Terpyridyl, see 2,6-di-2'-pyridylpyridine
   Thioacetamide, CH_3CSNH_2
   Thiourea, $SC(NH_2)_2$, tu
   2,2',2''-Triaminotriethylamine, tren (triam)
   Triethylamine oxide, $(C_2H_5)_3NO$
   Triethylarsine, $(C_2H_5)_3As$
   Triethylenetetramine, $(-CH_2NHCH_2CH_2NH_2)_2$, trien
   Triethylphosphine, $(C_2H_5)_3P$
   Triethylphosphine oxide, $(C_2H_5)_3PO$
   Triethylphosphite, $(C_2H_5O)_3P$
   Trimethylenediamine, see 1,3-propanediamine
   Urea, $OC(NH_2)_2$

C. **Negatively Charged (Anionic) Groups**

In general, if the name of the free anion ends in -ide, -ite, or -ate, the terminal -e is replaced by -o, giving the endings -ido, -ito, and -ato. However, there are a number of exceptions to this rule. Certain members of the -ide class have -ide replaced by -o—e.g., F^-, Cl^-, Br^-, I^-, O^{--}, OH^-, O_2^{--}, CN^-. Also, alternative names are sometimes used when a ligand can be bound through alternative atoms—e.g., M—NO_2 and M—ONO, nitro and nitrito. When naming compounds containing organic ligands, the definite replacement of hydrogen is indicated by the termination -ato.

1. Inorganic

   a. Singly charged groups

   Amido, NH_2^-
   Azido, N_3^-
   Bromato, BrO_3^-
   Bromo, Br^-
   Chlorato, ClO_3^-
   Chloro, Cl^-
   Cyanato, —OCN^-
   Cyano, —CN^-
   Fluoro, F^-
   Hydrazido, H_2NNH^-
   Hydrido, H^-
   Hydrocarbonato, HCO_3^-
   Hydroperoxo, HO_2^-
   Hydrosulfido, SH^-
   Hydroxo, OH^-
   Iodato, IO_3^-
   Iodo, I^-
   Isothiocyanato, —NCS^-
   Metaborato, BO_2^-
   Metaphosphato, PO_3^-
   Nitrato, NO_3^-
   Nitrito, —ONO^-
   Nitro, —NO_2^-
   Nitrosylo, NO^-
   Perchlorato, ClO_4^-
   Periodato, IO_4^-
   Selenocyanato, —$SeCN^-$
   Thiocyanato, —SCN^-

   b. Doubly charged groups

   Carbonato, CO_3^{--}
   Chromato, CrO_4^{--}
   Imido, NH^{--}
   Metasilicato, SiO_3^{--}
   Oxo, O^{--} (see Rule 15)
   Peroxo, O_2^{--}
   Selenato, SeO_4^{--}
   Selenido, Se^{--}
   Selenito, SeO_3^{--}
   Sulfamido, $SO_2(NH)_2^{--}$
   Sulfato, SO_4^{--}
   Sulfido, S^{--}
   Sulfito, SO_3^{--}
   (Thiosulfato), $S_2O_3^{--}$
   (Trithiocarbonato), CS_3^{--}

   c. Triply charged groups

   Nitrido, N^{---}
   Phosphato, PO_4^{---}

   d. Quadruply charged group

   Pyrophosphato, $P_2O_7^{----}$

Table I. List of Coordinated Groups (continued)

C. Negatively Charged (Anionic) Groups (continued)

2. Organic

Acetato (ethanoato), CH_3COO^-
Acetylacetonato, see 2,4-pentanedionato
2-Aminoethanethiolato, $H_2NCH_2CH_2S^-$
2-Aminoethylimido, $H_2NCH_2CH_2NH^-$
Benzoylacetonato, $PhCOCH=CO^-—CH_3$
Benzoylpyruvato, $C_6H_5COCH=CO^-—COO^-$
Catecholato
Dibenzoylmethanato, $PhCOCH=CO^-—Ph$
Diethylmalonato, $EtOCOCH—CO^-—OEt$
Dimethylamido, $(CH_3)_2N^-$
Dimethylglyoximato, $HON=CCH_3—CCH_3=NO^-$
Dithioöxalato, $SOC—COS^{--}$
Ethanethiolato, $C_2H_5S^-$
Ethoxido, $C_2H_5O^-$
Ethyl acetoacetato, $CH_3COCH_2=CO^-—COOC_2H_5$
α,α'-(Ethylenedinitrilo)di-o-cresolato, $(—CH_2N=CHC_6H_4O^-)_2$
4,4'-(Ethylenedinitrilo)di-2-pentene-2-olato, $(—CH_2N=CCH_3CH=CO^-CH_3)_2$
Ethyleneglycolato, $^-OCH_2CH_2O^-$
Ethylxanthato, $C_2H_5OCSS^-$
Formato, $HCOO^-$
Glycinato (aminoacetato, aminoethanoato), $H_2NCH_2COO^-$
Hydroöxalato, $HOOCCOO^-$
Malonato, $^-OOCCH_2COO^-$
Methanethiolato, CH_3S^-
Methoxido, CH_3O^-
Methyl, CH_3^-
1-Nitroso-2-naphtholato
Oxalato, $^-OOCCOO^-$, ox
Oxinato, see 8-quinolinolato
2,4-Pentanedionato, $CH_3COCH=CO^-—CH_3$
Phenolato, $C_6H_5O^-$
o-Phenylazophenolato, $C_6H_5N=NC_6H_4O^-$
α,α'-(o-Phenylenedinitrilo)di-o-cresolato
Phthalato
8-Quinolinolato
Salicylaldehydato
Salicylaldiminato
Salicylaldoximato
Salicylato
Succinimido
Thenoyltrifluoroacetonato
Thioglycolato, $^-SCH_2COO^-$
Xanthato, see ethylxanthato

Occasionally, modification of the name of the parent compound may be necessary in order to indicate the specific atoms through which chelation occurs.

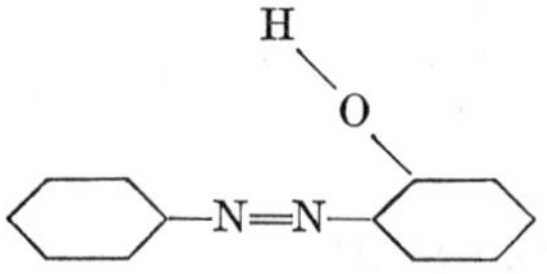

o-Hydroxyazobenzene

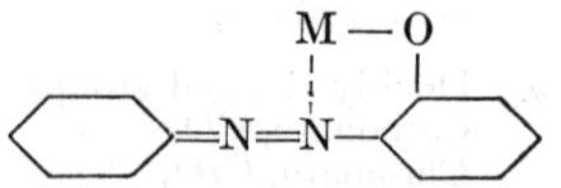

(o-Phenylazophenoxido)metal

the -o to negatively charged groups. At the Amsterdam meeting of the Commission on Inorganic Chemical Nomenclature of the International Union of Pure and Applied Chemistry (3), the use of the final -o for neutral groups was left optional. When used, it was supposed to emphasize the coordinate character of the linkage. However, all neutral groups are held by what must be essentially coordinate linkages; hence there is no need for the emphasis suggested. Such practice can lead only to confusion, particularly because the practice would be optional.

In order to include all groups frequently encountered as coordinating groups, it is desirable to compile a complete list and obtain general agreement on suitable names and abbreviations. The list in Table I is offered for this purpose.

Rule 3. Order of classes of ligands. The order of listing attached groups is (1) positive, (2) neutral, and (3) negative, without separation by hyphens. Reasons for the reversal of order from previous practices are given below. The use of parentheses (see Rule 5) is superior to the use of hyphens. Unless other considerations govern the particular case, uniformity and clarity would result by using the same order of attached groups in written formulas as in the name.

Examples:

[Co(NH$_3$)$_4$Cl$_2$]$^+$ Tetramminedichlorocobalt(III) ion
[Pt(H$_2$NNH$_3$)$_2$(NH$_3$)$_2$]$^{++++}$, 4Cl$^-$ Dihydraziniumdiammineplatinum(II) (tetra)chloride

COMMENT. It has always been the practice to list negative groups before neutral groups. However, Ewens and Bassett (7) present good arguments for reversing this order and their suggestions have been received favorably. In particular, the reversed order prevents confusion with an organic ligand having a negative substituent:

K$^+$, [Pt(C$_2$H$_4$)Cl$_3$]$^-$ Potassium ethylenetrichloroplatinate(II)

Trichloroethylene would suggest

$$\text{Cl}_2\text{C=CHCl}$$

[Pt(Et$_3$N)Cl$_3$]$^-$ Triethylaminetrichloroplatinate(II) ion

Trichlorotriethylamine would suggest (ClCH$_2$CH$_2$)$_3$N

The omission of hyphens is at variance with the tentative recommendations of the Commission on Inorganic Chemical Nomenclature of the International Union of Pure and Applied Chemistry (3) which states: "Long names should be broken up by means of hyphens if necessary if it improves the clarity." This rule is not satisfactory because:

1. The rules for the nomenclature of organic compounds do not provide for hyphens.
2. The use of parentheses (again following established practice in organic nomenclature) provides a better solution to the problem than hyphens.
3. A rule as indefinite as that of the commission can lead only to confusion and not to a standard practice.

Rule 4. Order within classes of ligands. The order of listing neutral groups is: (1) chelate groups (in order of decreasing complexity or number of atoms) and (2) simple groups; the order of listing negative groups is (1) O^{--}, OH$^-$ (in that order), (2) organic anions in order of decreasing complexity, (3) polynuclear inorganic anions, and (4) mononuclear anions (in the order of decreasing complexity: SO$_4^{--}$, CO$_3^{--}$, NO$_2^-$, O$_2^{--}$, OCl^{--}, Br$^-$, etc.).

Note. Within each of the subclasses of neutral groups the order follows the periodic table: B, Si, C, Sb, As, P, N, Te, Se, S, O, I, Br, Cl, F compounds with the more complex first whenever there are two members of the same type. This same order of elements is followed in the final breakdown with the last two subclasses of negative groups.

Examples:

[Fe(C$_5$H$_7$O$_2$)(OC$_2$H$_5$)$_2$]   2,4-Pentanedionatodiethoxidoiron(III)
Cs$^+$, [ICl$_2$F]$^-$ Cesium trichlorofluoroiodate(III)
Na[Co(C$_5$H$_7$O$_2$)$_2$(NO$_2$)$_2$]   Sodium bis(2,4-pentanedionato)dinitrocobaltate(III)
[Co(en)$_2$(NO$_2$)Cl]$^+$ Bis(ethylenediamine)nitrochlorocobalt(III) ion
2K$^+$, [NbOF$_5$]$^{--}$ Potassium oxopentafluoroniobate(V)
[Co(NH$_3$)$_5$H$_2$O]$^{+++}$ Pentammineaquacobalt(III) ion
[Pt(en)(NH$_3$)$_2$]$^{++}$ (Ethylenediamine)diammineplatinum(II) ion
[Pt(py)$_2$(NH$_3$)$_2$]$^{++}$ Dipyridinediammineplatinum(II) ion
[Pt(C$_2$H$_4$)NH$_3$Cl$_2$]   (Ethylene)amminedichloroplatinum(II)

COMMENT. This rule embodies essentially the recommendations of Ewens and Bassett (7) and is in harmony with the order of elements recommended by the IUPAC

(*3*) for citing anions in mixed and double salts. A detailed order for cationic substituents is unnecessary at this stage.

Rule 5. Use of numerical prefixes. The prefixes **di-, tri-, tetra-,** etc., are used before simple expressions and the prefixes **bis-, tris-, tetrakis-,** etc., before complex expressions. All complex expressions are enclosed in parentheses.

Examples:

[Cr(en)$_3$]$^{+++}$ — Tris(ethylenediamine)chromium(III) ion
[Pt{OP(CH$_3$)$_3$}$_2$(NH$_3$)$_2$]$^{++}$, 2Cl$^-$ — Bis(trimethylphosphine oxide)diammineplatinum(II) chloride

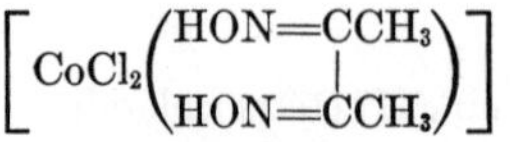

$$\left[CoCl_2 \begin{pmatrix} HON=CCH_3 \\ | \\ HON=CCH_3 \end{pmatrix} \right]$$
(Dimethylglyoxime)dichlorocobalt(II)

Note. The prefix **bis** has been used previously to indicate the coordination of doubled molecules:

[Be(H$_4$O$_2$)$_4$]$^{++}$, [PtCl$_6$]$^{--}$ — Tetra**bis**aquaberyllium hexachloroplatinate(VI)
[Co(H$_2$O$_2$)$_6$]$^{++}$, S$_2$O$_3$$^{--}$ — Hexa**bis**aquacobalt(II) thiosulfate

Although the coordination of doubled molecules is still open to question, the prefix **bi** could be used in this connection to avoid any confusion with the use of **bis** as prescribed above:

Tetrabiaquaberyllium hexachloroplatinate(VI)
Hexabiaquacobalt(II) thiosulfate

COMMENT. Rule 67 of the Definitive Report of the Commission on the Reform of the Nomenclature of Organic Chemistry (*28*) reads:

The prefixes, **di, tri, tetra,** etc., will be used before simple expressions (for example, diethylbutanetriol) and the prefixes **bis, tris, tetrakis,** etc., before complex expressions. Examples: bis(methylamino)propane: CH$_3$NH(CH$_2$)$_3$NHCH$_3$; bis(dimethylamino)ethane, (CH$_3$)$_2$NCH$_2$CH$_2$N(CH$_3$)$_2$ . . . . .

Since this rule has served very well for organic nomenclature, it seems wise to extend it to corresponding use in the inorganic field. The extent to which this has already happened indicates the practicability of the rule. Further, the practice of using parentheses to enclose complex expressions eliminates any use of hyphens to "improve clarity" (*3*) (see COMMENT under Rule 3).

Note. When the ligand with complex name occurs only once, it may be enclosed in parentheses where otherwise there might be confusion concerning the structure of the compound:

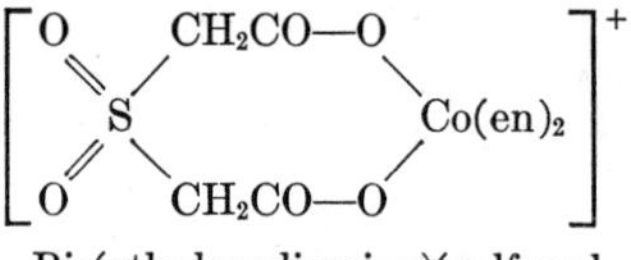

Bis(ethylenediamine)(sulfonyldiacetato)cobalt(III) ion

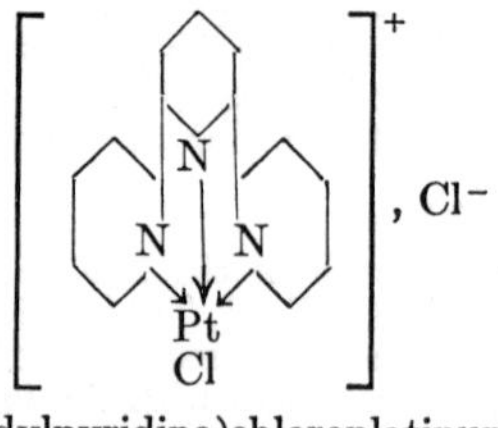

(2,6-Di-2'-pyridylpyridine)chloroplatinum(II) chloride
or (α,α',α''-Terpyridine)chloroplatinum(II) chloride

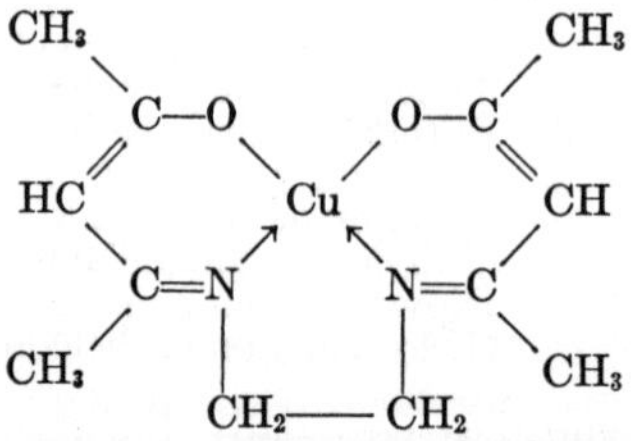

{4,4'-(Ethylenedinitrilo)di-2-pentene-2-olato}copper(II)

Rule 6. Terminations for anions. The characteristic termination for an anionic coordination entity is **-ate,** or **-ic** if named as an acid. There are no characteristic terminations for cationic or neutral coordination entities.

Examples:

$3K^+$, $[Co(NO_2)_6]^{---}$	Potassium hexanitrocobaltate(III)
$H_4[Fe(CN)_6]$	Hydrogen hexacyanoferrate(II)
$H_3[Mn(CN)_6]$	Hexacyanomanganic(III) acid
$[Cr(H_2O)_6]^{+++}$, $3Cl^-$	Hexaquachromium(III) chloride
$[Cr_3(OH)_2(CH_3CO_2)_6]^+$, X^-	Dihydroxohexacetatotrichromium(III) halide
$[Ni(C_5H_7O_2)_2]$	Bis(2,4-pentanedionato)nickel(II)

COMMENT. Again this rule is largely a recognition of what has been standard practice. If $H_4[Mo(CN)_8]$ is named as an acid instead of a hydrogen salt, then it is consistent with standard practice to replace the suffix **-ate** by **-ic**:

Hydrogen octacyanomolybdate(IV)
Octacyanomolybdic(IV) acid

Rule 7. Designation of oxidation state. The oxidation state of the central element is designated by a Roman numeral in parentheses: for a cationic or neutral coordination entity, after the name of the central element; for an anionic entity, after the termination, -ate.

Note. The Arabic 0 is used for zero.

Examples:

$[Ru(NH_3)_4Cl_2]^+$, Cl^-	Tetramminedichlororuthenium(III) chloride
$[Ir(NH_3)_3(NO_2)_3]$	Triamminetrinitroiridium(III)
K^+, $[Au(OH)_4]^-$	Potassium tetrahydroxoaurate(III)
$4K^+$, $[Ni(CN)_4]^{----}$	Potassium tetracyanonickelate(0)
$[Rh(NH_3)_6]^{+++}$, $3NO_3^-$	Hexamminerhodium(III) nitrate
$4K^+$, $[Os(CN)_6]^{----}$	Potassium hexacyanoösmate(II)

COMMENT. This rule is simply an extension of the Stock system to coordination compounds.

Ewens and Bassett (7) have presented arguments favoring the indication of ionic charge (in Arabic numerals) instead of the oxidation state. They state that the expression "zero valent elements" in such compounds as $Ni(CO)_4$ and $K_4[Ni(CN)_4]$ is meaningless. It is unfortunate that the terms valence and valency have been used in the sense both of oxidation state and of number of bonds. The expression "element of oxidation state zero" is not meaningless but indicates the most important characteristic of these compounds.

In support of continuing the practice of indicating the oxidation state of the central element are the following:

1. It has long been customary to make classifications of our knowledge of chemical compounds on the basis of oxidation state. There seems to be little likelihood of any change in this practice.

2. The thing in common between a free ion and a coordination entity derived from it is the oxidation state. Comparisons and differences between the physical properties (such as magnetic susceptibility) of the two are made on the basis of oxidation state. These relationships would be obscured by any change from the present practice.

While there are a few cases, such as $K_2[Fe_2(NO)_4S_2]$ and $K[Fe_4(NO)_7S_3]$, in which the state of oxidation of the central atom is in doubt, these are not numerous and unquestionably will be resolved by future work. If the need for resolution of a nomenclature problem stimulates laboratory investigation, the science of chemistry is enriched.

3. The practice of using Roman numerals has become widespread. Any change at this time would add still another nomenclature system to a field which is already confused by too many different systems. Consequently, there ought to be very convincing reasons for making any change.

Rule 8. Bridging groups. Those groups which act to bridge two centers of coordination are designated by the Greek letter μ. In case more than one kind of bridging group is present in the coordination entity, the letter μ is repeated before the name of each bridging group.

Note. The order of listing bridging groups cannot always follow that given in Rule 3.

Examples:

[(NH$_3$)$_5$Cr—OH—Cr(NH$_3$)$_5$] +++++ μ-Hydroxobis{pentamminechromium(III)} ion or decammine-μ-hydroxodichromium(III) ion

[(CS$_2$)(NH$_3$)$_3$Co—S—Co(CS$_2$)(NH$_3$)$_2$] μ-Sulfidobis{triammine(trithiocarbonato)cobalt(III)}, triammine(trithiocarbonato)-cobalt(III)-μ-sulfidotriammine(trithiocarbonato)cobalt(III), or bis{triammine-(trithiocarbonato)cobalt(III)}sulfide

[(NH$_3$)$_5$Fe—CN—Co(CN)$_5$] Pentamminiron(III)-μ-cyanopentacyanocobalt(III)

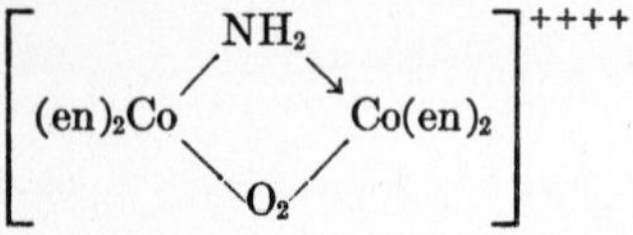

$\left[(en)_2Co \begin{smallmatrix} NH_2 \\ \diagup \diagdown \\ O_2 \end{smallmatrix} Co(en)_2 \right]^{++++}$ Tetrakis(ethylenediamine)-μ-peroxo-μ-amidodicobalt(III, IV) ion or bis(ethylenediamine)cobalt(III)-μ-peroxo-μ-amidobis(ethylenediamine)cobalt(IV) ion

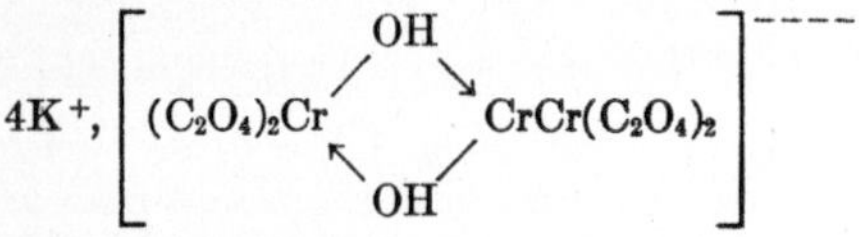

$4K^+, \left[(C_2O_4)_2Cr \begin{smallmatrix} OH \\ \diagup \diagdown \\ OH \end{smallmatrix} CrCr(C_2O_4)_2 \right]^{----}$ Potassium tetraoxalato-μ-dihydroxodichromate(III)

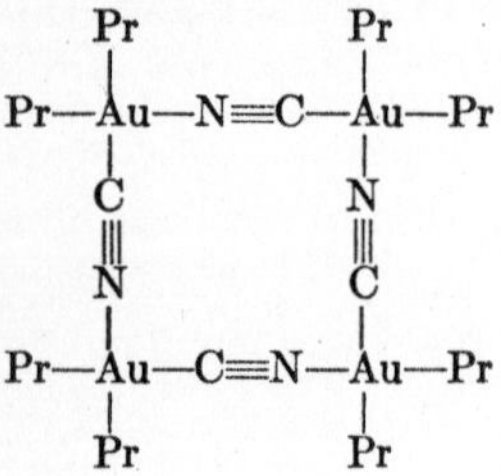

Pr Pr
| |
Pr—Au—N≡C—Au—Pr
| |
C   N
||| |||
N   C
| |
Pr—Au—C≡N—Au—Pr
| |
Pr Pr

μ-Tetracyanotetrakis(dipropylgold), tetrakis-(dipropyl-μ-cyanogold), or dipropylgold cyanide tetramer

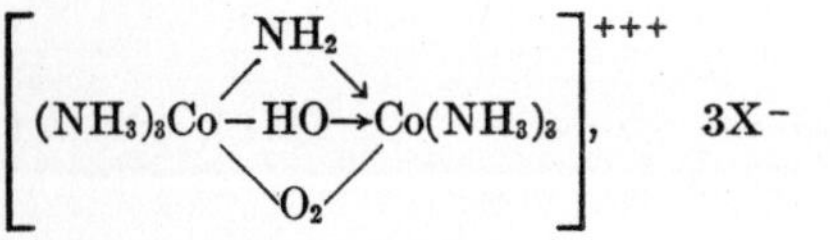

$\left[(NH_3)_3Co \begin{smallmatrix} NH_2 \\ \diagup \diagdown \\ HO \to \\ \diagdown \diagup \\ O_2 \end{smallmatrix} Co(NH_3)_2 \right]^{+++}, 3X^-$ Triamminecobalt(III)-μ-hydroxo-μ-amido-μ-peroxotriamminecobalt(IV) trihalide or hexammine-μ-hydroxo-μ-amido-μ-peroxodicobalt(III, IV) trihalide

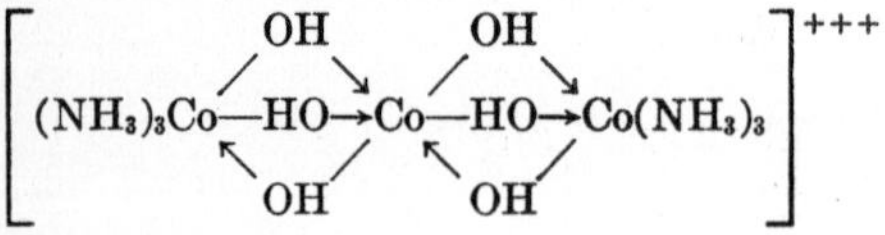

$\left[(NH_3)_3Co \begin{smallmatrix} OH & OH \\ HO\to Co\to HO \\ OH & OH \end{smallmatrix} Co(NH_3)_3 \right]^{+++}$ Bis{triammine-μ-trihydroxocobalt(III)}-cobalt(III) ion or triamminecobalt(III)-μ-trihydroxocobalt(III)-μ-trihydroxotriamminecobalt(III) ion

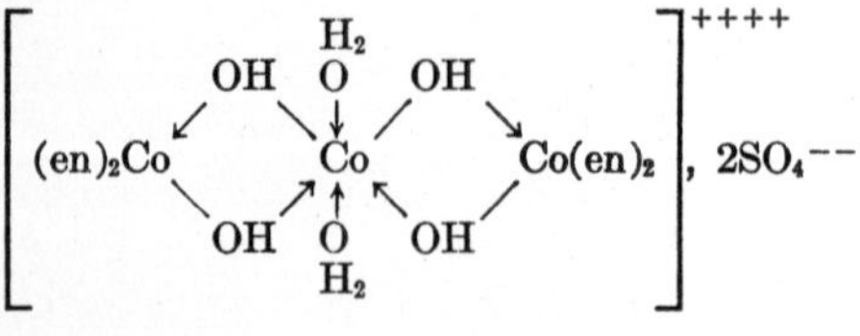

$\left[(en)_2Co \begin{smallmatrix} & H_2 & \\ OH & O & OH \\ & Co & \\ OH & O & OH \\ & H_2 & \end{smallmatrix} Co(en)_2 \right]^{++++}, 2SO_4^{--}$ Bis{bis(ethylenediamine)-μ-dihydroxocobaltium(III)}diaquacobalt(II) sulfate

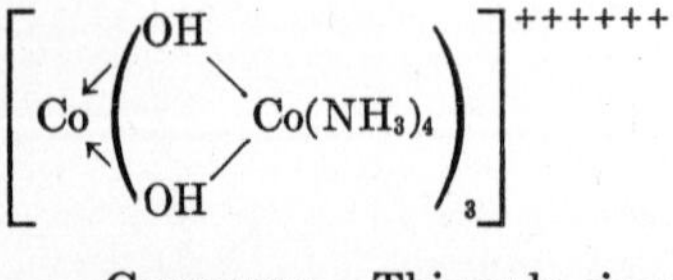

$\left[Co \left(\begin{smallmatrix} OH \\ \diagdown \\ \diagup \\ OH \end{smallmatrix} Co(NH_3)_4 \right)_3 \right]^{++++++}$ Tris{tetrammine-μ-dihydroxocobaltium(III)}cobalt(III) ion or tris{tetramminedihydroxocobaltium(III)}cobalt(III) ion

COMMENT. This rule simply codifies a practice of long standing. It has been customary to use the term **-ol** in place of μ-**hydroxo** when a hydroxyl group acts as a bridging group. While this type of bridging is probably common, the special designation at variance with practice has not been widely used.

Note. While most bridging groups are bifunctional, there are some of higher function. This rule may need modification to cover such cases. Consider the substance:

$$(C_4H_9)_3P \diagdown Pd \diagup Cl \diagup \text{(oxalato bridges)} \diagdown Pd \diagup P(C_4H_9)_3$$

The structure would not be evident unless the functionality of the bridging group were indicated in some manner such as:

Bis(tripropylphosphine)dichloro-μ^4-oxalatodipalladium(II) or
μ^4-oxalatobis{(tributylphosphine)chloropalladium(II)}

This principle also makes possible a descriptive name for the troublesome compound known as "basic beryllium acetate," $Be_4O(CH_3CO_2)_6$ (*2, 23, 30*):

μ^4-Oxo-μ-hexaacetatotetraberyllium

Rule 9. Designation of the point of attachment. Whenever it is desired to designate the point of attachment, this is done by placing the symbols (in italics) of the elements through which coordination occurs after the name of the coordinated group.

Examples:

Bis(dimethylglyoximato-N,N')palladium(II)

Potassium bis(dithiooxalato-S,S')nickelate(II)

Note. Further examples of groups which frequently may require that the point of attachment be designated are the following:

$C_6H_5NHNHCS-N=NC_6H_5 \rightleftarrows C_6H_5NH-N=C(SH)-N=NC_6H_5$

The Greek letters, α, β, γ, etc., may be used to designate the position on the carbon chain of the elements through which coordination occurs:

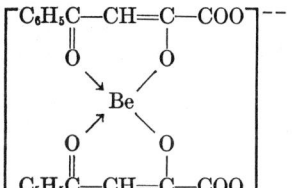

Potassium bis(benzoylpyruvato-O^α,O^γ)beryllate

This type of designation is particularly useful when, as in the instance above, there are several possible atoms of the same element through which coordination may occur, but coordination actually occurs through only a part of them. In case the point of attachment is not definitely established, it may be desirable to express its uncertainty by the use of a question mark. The metal derivatives of dithizone (diphenyl thiocarbazone) and of rubeanic acid (dithioöxamide) are instances where the structures are uncertain.

COMMENT. There is often a real need to designate the point or points of attachment of a coordinated group. This is desirable to indicate isomerism (see Rule 10). In cases where only one of two or more hypothetical isomers seems possible, it is helpful to indicate the metal-atom linkage because certain properties are often characteristic of specific linkages.

The symbols are placed **after** the name of the coordinating group in order to avoid any confusion with the use of such symbols in organic nomenclature. Consider the following hypothetical but by no means unlikely compounds:

1.
$$\begin{bmatrix} Cl & & H_2NCH_2 \\ & \diagdown Pt \diagup & \\ & \diagup \diagdown & \\ Cl & & S\!-\!CH_2 \\ & & | \\ & & CH_2CH_2N(CH_3)_2 \end{bmatrix}$$

2. $[Pt\{SP(OCH_3)_2(SCH_3)\}_2(NH_3)_2]^{++}$, $2Cl^-$
3. $[Pt\{OP(SCH_3)_2(OCH_3)\}_2(NH_3)_2]^{++}$, $2Cl^-$

The name (N,N-dimethyldiaminodiethylsulfide-N',S)dichloroplatinum(II) for (1) indicates clearly that the sulfur-containing amine is coordinated to the platinum by the S-atom and by the unsubstituted N-atom. The names bis(O,O',S-trimethyldithiophosphate-S')diammineplatinum(II) chloride and bis(O,S,S'-trimethyldithiophosphate-O')diammineplatinum(II) chloride for (2) and (3) describe each completely and thus make the differences between them clear.

Note. Traditionally, element symbols have been placed before the name of the substituted group. Consequently, many who have read the manuscript of this article have tried to devise ways of placing the symbols designating atoms through which coordination takes place before the name of the coordinated group. The most successful of these attempts yields the following names for the examples given under Rule 9:

Bis-N,N'-(dimethylglyoximato)palladium(II)
Potassium bis{S,S'-(dithioöxalato)}nickelate
N',S-{N,N-dimethyl(2,2'-thiobisethylamine)}dichloroplatinum(II)
Bis-S'-(O,O',S-trimethyldithiophosphate)diammineplatinum(II) chloride
Bis-O'-(O,S,S'-trimethyldithiophosphate)diammineplatinum(II) chloride

The author still prefers a complete separation of symbols designating atoms through which coordination takes place from those designating atoms on which there is substitution for hydrogen. Others may feel differently.

Rule 10. Designation of structural isomerism. Structural isomerism is designated by the use of different terms for the stoichiometrically equivalent groups or by the method of Rule 9. The following terms for isomeric groups is a practice of long standing:

—NO_2 Nitro-   —SCN Thiocyanato-

—ONO Nitrito-   —NCS Isothiocyanato-

For the future a preferable alternate to the coining of new terms for the same group when coordinated through different kinds of atoms is the method of Rule 9 as in the following:

$$\begin{array}{c} -O \quad\ \ S \\ \diagdown\!\!\diagup \\ S \\ \diagup\!\!\diagdown \\ -O \quad\ \ O \end{array} \quad \text{Thiosulfato-}O,O- \qquad\qquad \begin{array}{c} -S \quad\ \ O \\ \diagdown\!\!\diagup \\ S \\ \diagup\!\!\diagdown \\ -O \quad\ \ O \end{array} \quad \text{Thiosulfato-}O,S-$$

Examples:

$\left[\text{Co}\begin{array}{c}\text{ONO}\\(\text{NH}_3)_5\end{array}\right]^{++}$ Pentamminenitritocobalt(III) ion

$\left[\text{Co}\begin{array}{c}\text{NO}_2\\(\text{NH}_3)_5\end{array}\right]^{++}$ Pentamminenitrocobalt(III) ion

$\left[\text{Co}\begin{array}{c}\text{NCS}\\(\text{NH}_3)_5\end{array}\right]^{++}$ Pentammineisothiocyanatocobalt(III) ion

$\left[\text{Cr}\begin{array}{c}\text{SCN}\\(\text{NH}_3)_5\end{array}\right]^{++}$ Pentamminethiocyanatochromium(III) ion

$\left[\text{Co}\begin{array}{c}\text{Br}\\\text{NCS}\\(\text{en})_2\end{array}\right]^{+}$ Bis(ethylenediamine)isothiocyanatobromocobalt(III) ion

Rule 11. Designation of structural isomerism. Geometrical isomerism is designated alternatively by numbers or by the words *cis-* and *trans-*.

A. The numbering of planar complexes is based on the following pattern:

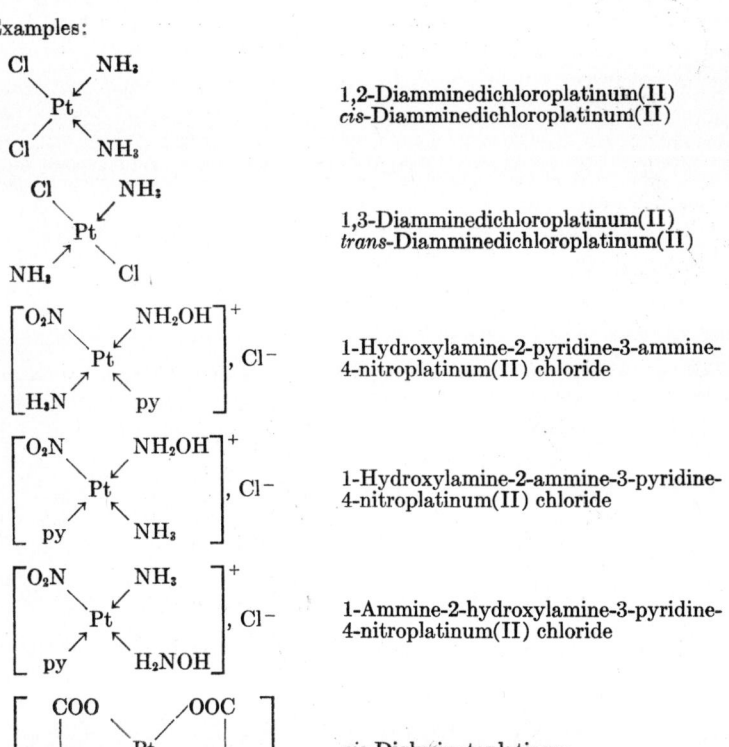

Examples:

1,2-Diamminedichloroplatinum(II)
cis-Diamminedichloroplatinum(II)

1,3-Diamminedichloroplatinum(II)
trans-Diamminedichloroplatinum(II)

1-Hydroxylamine-2-pyridine-3-ammine-4-nitroplatinum(II) chloride

1-Hydroxylamine-2-ammine-3-pyridine-4-nitroplatinum(II) chloride

1-Ammine-2-hydroxylamine-3-pyridine-4-nitroplatinum(II) chloride

cis-Diglycinatoplatinum

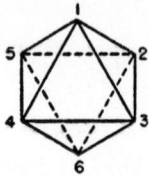

trans-Diglycinatoplatinum

B. The numbering of octahedral complexes is based on the following pattern:

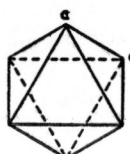

1. For entities of the type [Ma₂b₄] there are two isomers:

1,2- or *cis*-   1,6- or *trans*-

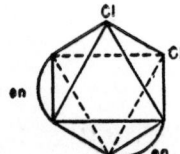

Bis(ethylenediamine)-1,2-dichlorocobalt(III) ion
cis-Bis(ethylenediamine)dichlorocobalt(III) ion

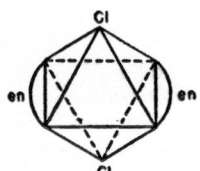

Bis(ethylenediamine)-1,6-dichlorocobalt(III) ion
trans-Bis(ethylenediamine)dichlorocobalt(III) ion

2. For entities of the type [Ma₃b₃], there are two possible isomers:

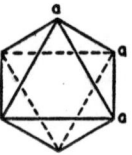

1,2,3- or *cis*-   1,2,6- or *trans*-

C. The numbering of binucleate complexes with two bridges is based on the following pattern:

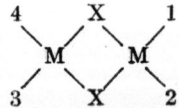

1. For planar entities of the type $(Mx_2M)a_2b_2$ there are three possible isomers:

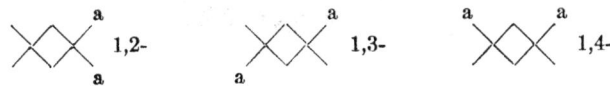

The following formulas are illustrative although they do not represent known compounds:

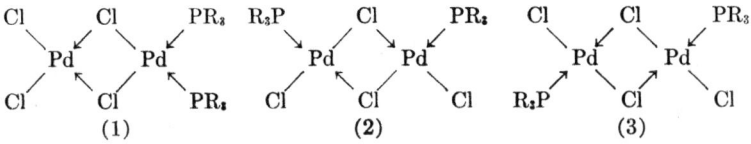

1. 1,2-Bis(triethylphosphine)-3,4-dichloro-μ-dichlorodipalladium
2. 1,4-Bis(triethylphosphine)-2,3-dichloro-μ-dichlorodipalladium
3. 1,3-Bis(triethylphosphine)-2,4-dichloro-μ-dichlorodipalladium

2. There would be no isomers for a symmetrical planar chelate compound with two bridges of the type $[(Mx_2M)(AA)_2]$ and two isomers for an unsymmetrical planar chelate compound of the type $[(Mx_2M)(AB)_2]$:

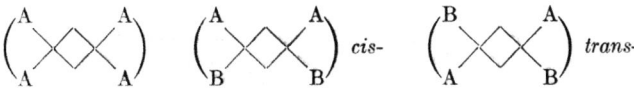

3. When the central atoms linked through bridges exhibit a tetrahedral distribution of bonds rather than a planar one, the following numbering pattern applies:

The shading of lines indicates that 1 and 4 are above the plane of the paper and 2 and 3 are below while the two M's and X's are in the plane. For entities of the type $[(Mx_2M)a_2b_2]$ there are again three isomers:

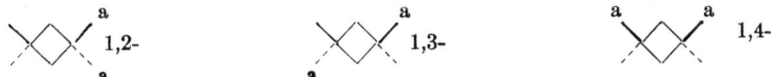

COMMENT. This rule recognizes standard practice. Moreover, it shows how this practice may be extended to types of compounds for which it has not hitherto been used.

Rule 12. Designation of optical isomerism. Where optical isomerism can occur, the optically active compound is designated by (+) or (−) depending upon the sign of rotation; alternatively, d- or l- may be used. The racemic mixture is designated by (±) or dl- and the inactive form by *meso*.

Note. The small capitals D and L are reserved for the configurational relationships of organic compounds in the optically active series (51).

Examples:

(+)-[Fe(C_{12}N_2H_8)_3]^{++}
(±)-[Co(en)_2C_2O_4]^{+}

(+)-Tris(*o*-phenanthroline)iron(II) ion
d-Tris(*o*-phenanthroline)iron(II) ion
(±)-Bis(ethylenediamine)oxalatocobalt(III) ion

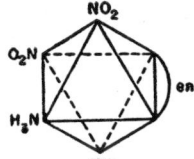

(−)-1,2-(Ethylenediamine)-3,4-diammine-5,6-dinitrocobaltate(III) ion
(−)-*cis*-(Ethylenediamine)diamminedinitrocobaltate(III) ion

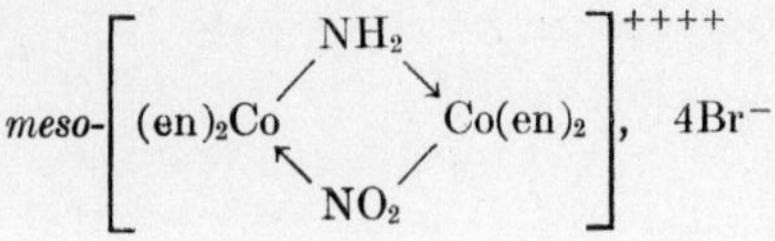

$$\left[meso\text{-} (en)_2Co \begin{array}{c} NH_2 \\ \diagup \diagdown \\ \diagdown \diagup \\ NO_2 \end{array} Co(en)_2 \right]^{++++}, \; 4Br^- \qquad meso\text{-Tetrakis(ethylenediamine)-}\mu\text{-amido-}\mu\text{-nitro-dicobalt(III) bromide}$$

A. The order of specifying the types of isomerism is (1) optical isomerism and (2) geometrical isomerism:

Example:

$(+)$-1,2-$[Co(en)_2Cl_2]^+$, Cl^- $\qquad$ $(+)$-Bis(ethylenediamine)-1,2-dichlorocobalt(III) chloride

B. In cases where there are optically active (or *meso*) ligands, these are treated as complex names:

Examples:

$(+)$-$[Rh\{(-)\text{-cpd}\}_3]^{+++}$, $3Cl^-$ $\qquad$ $(+)$-Tris$\{(-)$-cyclopentanediamine$\}$rhodium(III) chloride
$[Pt(\text{D-alanine})_2]$ $\qquad\qquad\qquad\qquad$ Di-D-alaninatoplatinum(II)
$(-)$-$[Pt\{meso\text{-}H_2N(CHC_6H_5)_2NH_2\}\{H_2NCH_2C(CH_3)_2NH_2\}]^{++}$
$\qquad\qquad\qquad\qquad\qquad\qquad (-)$-(*meso*-Stilbenediamine)(1,2-diamino-2-methylpropane)platinum(II) ion

NOTE. Inasmuch as marked rotary dispersion is encountered with many optically active coordination compounds, it is advisable to designate the wave length of light at which the rotation was measured. This is especially important because the **sign** of rotation actually reverses within the visible spectrum for many coordination compounds.

It is also well to designate the concentration at which the rotation is measured because many compounds undergo a change in sign of rotation with a change in concentration.

COMMENT. This rule is simply an attempt to adapt the best organic nomenclature practices to coordination compounds. Difficulty arises because those who have worked with optically active inorganic compounds almost invariably have used d- and l- to designate the sign of rotation of the coordinated group and large capitals D and L for the sign of rotation of the coordination entity.

Rule 13. Coordination of groups in lower functionality than usual. The nomenclature of compounds involving monocoordination of bifunctional groups and the dicoordination of polyfunctional groups, etc., usually can be handled according to the rules previously given.

Examples:

For the compound represented by the formula $[Co(NH_3)_5SO_4]Br$ the name pentamminesulfatocobalt(III) bromide indicates clearly that the sulfate group is monocoordinated if a maximum coordination number of six is accepted for cobalt.

The above statement applies to the name pentammineoxalatocobalt(III) sulfate for the compound represented by the formula $[Co(NH_3)_5C_2O_4]_2SO_4$.

For the compound represented by the formula:

$$\left[\begin{array}{cc} Cl & NH_2\text{---}CH_2 \\ & \quad\quad\quad\quad | \\ Pt & \\ & \quad\quad\quad\quad | \\ Cl & NH_2\text{---}CH \\ & \quad\quad\quad\quad | \\ & \quad\quad\quad CH_2NH_3 \end{array} \right]^+, \; Cl^-$$

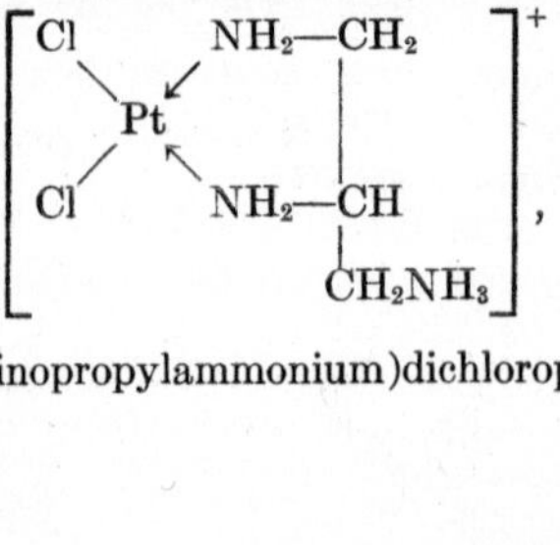

the name might be (2,3-diaminopropylammonium)dichloroplatinum(II) chloride.

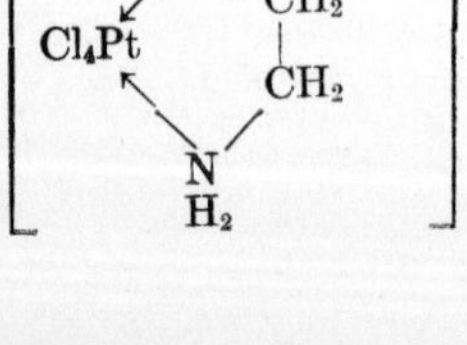

$$\left[Cl_4Pt \begin{array}{c} C_2H_4NH_2 \\ \diagup \\ S \\ | \\ CH_2 \\ | \\ CH_2 \\ | \\ N \\ \diagdown \\ H_2 \end{array} \right] \qquad \{\text{Bis(2-aminoethyl)sulfide-}N,S\}\text{tetrachloroplatinum(IV)}$$

$$\begin{bmatrix} \text{NH}_2 & & \text{NH}_2 \\ | & & | \\ \text{CH}_2 & \searrow \text{Pt} \swarrow & \text{CH}_2 \\ | & \diagup \quad \diagdown & | \\ \text{CH}_2 & \text{Cl} \quad \text{Cl} & \text{CH}_2 \\ | & & | \\ \text{NH}_3 & & \text{NH}_3 \end{bmatrix}^{++}, \; 2\text{Cl}^-$$

Bis(2-aminoethylammonium)dichloroplatinum(II) chloride

Rule 14. Direct linking of coordination centers. The direct linking of two centers of coordination is designated by the prefix **bi-** before the name of the coordination centers.

Example:

$$\begin{bmatrix} (\text{EtNH}_2)_4\text{Pt} - \text{Pt}(\text{EtNH}_2)_4 \\ \quad \quad \quad | \quad | \\ \quad \quad \quad \text{Cl} \; \text{Cl} \end{bmatrix}^{++++}$$

sym-Octakis(ethylamine)dichlorobiplatinum(IV) ion

COMMENT. The reality of the Pt-Pt linkage is still in doubt but this pattern of nomenclature may have applications in another connection (see treatment of carbonyl compounds later).

Rule 15. Coordinated oxide ions. Since oxide ions are best regarded as occupying specific coordination positions, they should be designated in the name rather than to use -yl to designate oxy ions as centers in coordination compounds.

Examples:

$\text{UO}_2(\text{C}_5\text{H}_7\text{O}_2)_2$ Dioxobis(acetylacetonato)uranium(VI) instead of uranyl acetylacetonate

Dioxo-{α,α'-(*o*-phenylenedinitrilo)di-*o*-cresolato}-uranium(VI)

Oxo-{α,α'-(ethylenedinitrilo)di-*o*-cresolato}vanadium(IV)

COMMENT. A further advantage of this nomenclature is that, as in the case of that of uranium above, it immediately points to the possible existence of isomeric forms of the composition represented by the formula.

Rule 16. Abbreviations in formulas. In writing the formulas of coordination compounds, it is frequently customary for convenience to use simple abbreviations consisting of two or more letters for complicated molecules.

COMMENT. Except in a few cases: en, ethylenediamine; pn, propylenediamine; and py, pyridine, there seems to be no general agreement among various authors as to the abbreviations for a particular group. Hence, in every case the significance of an abbreviation should be made clear.

In order that as much agreement as possible be obtained, a list of abbreviations which have been used by various authors is given in Table I where the names of various types of ligands are listed.

In writing formulas care should be taken to avoid ambiguity in the use of abbreviations.

Of the three methods:

$[Co\ en_3]^{+++}$
$[Co3en]^{+++}$
$[Co(en)_3]^{+++}$

the second is confusing while the last cannot lead to misinterpretation.

While some authors have capitalized these abbreviations in harmony with Me, Et, etc., the majority have not used capital letters.

Applications of the Proposed Rules

Any scheme of nomenclature for coordination compounds should offer a broad general pattern capable of wide extension. These rules provide suitable names for compounds to which this pattern of nomenclature has not been applied previously.

The table below gives a number of related compounds together with suggested names.

1 $Na^+, [Al(OH)_4]^-$ — Sodium tetrahydroxoaluminate
2 $Na^+, [Al(OC_2H_5)_4]^-$ — Sodium tetraethoxidoaluminate
3 $Na^+, [B(OCH_3)_3H]^-$ — Sodium triethoxidohydridoborate
4 $Na^+, [Al(NH_2)_4]^-$ — Sodium tetramidoaluminate
5 $Li^+, [Al(NMe_2)_4]^-$ — Lithium tetrakis(dimethylamido)aluminate
6 $Li^+, [Al(NH_2)_3H]^-$ — Lithium triamidohydridoaluminate
7 $2Li^+, [\{AlH(NH_2)_2\}_2NH]^{--}$ — Lithium tetramidodihydrido-μ-imidodialuminate
8 $Li^+, [AlH_4]^-$ — Lithium tetrahydridoaluminate
9 $2Li^+, [Zn(CH_3)_4]^{--}$ — Lithium tetramethylzincate
10 $Li^+, [B(CH_3)_3H]^-$ — Lithium trimethylhydridoborate

The first of these represents no marked departure from practice but the second, which follows logically from the first, seems not to have been used.

The names sodium borohydride and lithium aluminum hydride (8) are not constructed on any established nomenclature pattern unless it be rapidly disappearing one represented by sodium silicofluoride and potassium ferrocyanide.

The third name above certainly is more likely to create a visual picture of the structure than the currently used "sodium trimethoxyborohydride" (40).

Similar intermediate hydrides are represented by examples 6 and 7 above (9). The names suggested show clearly their relationship to 4, 5, and 8.

Any pattern of nomenclature for the complex hydrides should be capable of extension to the complex alkyl and aryl derivatives. The last two examples indicate how this is accomplished. In strict analogy the alkyl groups should be designated methylo, ethylo, etc., but this may depart too much from custom and imply too definite a conviction as to the nature of the carbon-metal bond.

These suggestions may be extended to the bridge hydrides and metal alkyls:

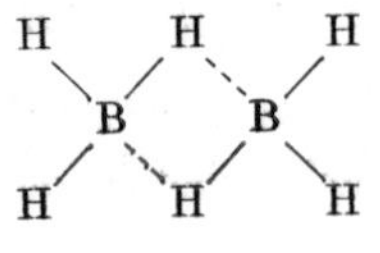

Tetrahydrido-μ-dihydridodiboron

$Al(BH_4)_3$ — Tris(tetrahydridoborato)aluminum

$(CH_3)_2GaBH_4$ — Tetrahydridoboratodimethylgallium

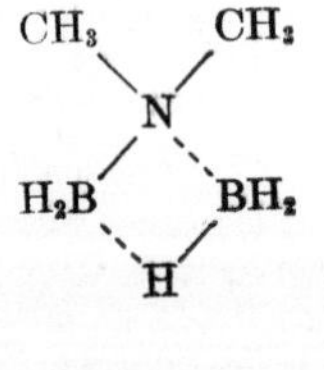

Tetrahydrido-μ-(dimethylamido)-μ-hydridodiboron

[Structure: Tetramethyl-μ-dimethyldialuminum]

$$\begin{array}{c} H_3C \\ H_3C \end{array} Al \begin{array}{c} CH_3 \\ \vdots \\ C \\ H_3 \end{array} Al \begin{array}{c} CH_3 \\ CH_3 \end{array}$$

Tetramethyl-μ-dimethyldialuminum

Other interesting cases of coordination compounds derived from organometallic compounds are the following:

[Structure] Tetraethyl-μ-dibromodigold

[Structure] (Ethylenediamine)diethylgold bromide

[Structure] sym-Tetrapropyldibromo-μ-(ethylenediamine)-digold

[Structure] 2,4-Pentanedionatodimethylthallium

[Structure] (Ethylenediamine)pentamethylenegold bromide

In the nomenclature suggestions up to this point, it is possible to differentiate among monodentate groups with positive, zero, or negative charges. This practice is extended to bidentate or multidentate groups with similar charges. However, no means has been suggested for differentiating between a bidentate group with a single negative charge and one with a double negative charge. This would be of importance in the case of substances such as glycol, pyrocatechol, o,o'-dihydroxyazo dyes, and similar substances which can release one or two protons. It seems reasonable that this problem can be solved in one of two ways: (1) modify the name of the ligand group to indicate displacement of hydrogen in one part of the molecule but not in another and (2) use symbols to indicate the points of attachment.

The examples presented by the o,o'-dihydroxyazo compounds can readily be handled by the first of the above methods. The examples given below demonstrate the method for dealing with simple substituted azo compounds and then extends the system to the troublesome case in which both protons are displaced from one ligand molecule and only one from the other.

[Structure] o-Hydroxyazobenzene, o-phenylazophenol

Bis(*o*-phenylazophenolato)copper(II)

Bis{2-(4'-chlorophenylazo)-5-nitrophenolato}nickel(II)

Bis(*o*-phenylazobenzoato)copper(II)

Bis(1-phenyl-3-methyl-4-phenylazo-5-pyrazolonato)zinc

Aqua{*o*,*o*'-azobis(phenolato)}copper(II)

Aqua{1-(2'-azophenolato)-2-naphtholato}lead

o,*o*'-Azobis(phenoxido){*o*-(2-hydroxyphenylazo)phenoxido}cobalt(III)

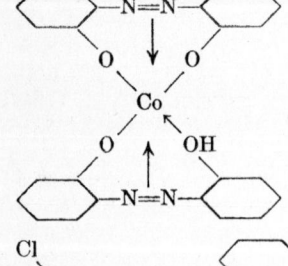

Bis{1-(5'-chloro-2'-azophenolato)-2-naphthylamine}cobalt(II)

Aqua{1-phenyl-3-methyl-4-(2'-azophenolato)-5-pyrazolonato}manganese(II)

These names are long, but then the names of most complicated molecules are long if they are truly descriptive.

Another suitable group of compounds for testing the adaptability of the nomenclature patterns for coordination compounds is that presented by the aminopolyacids. The metal derivatives of some of these are of great interest today; so much so, in fact, that ethylenediaminetetraacetic acid has become an important article of commerce. There have been, as yet, no serious efforts to name the metal coordination compounds of these acids. The following are suggested:

$H_4[Cu\{N(CH_2COO)_3\}_2]$
    4-Coordinate
    Bis(nitrilodiacetate acetato-N,O)copper(II) acid

$H_4[Cu\{N(CH_2COO)_3\}_2]$
    6-Coordinate
    Hydrogen bis(nitriloacetate diacetato-N,O,O')-copperate(II)

$[Cr(H_2O)_2N(CH_2CO_2)_3]$
    Diaqua(nitrilotriacetato-N,O,O',O'')chromium(III)

$H_6[Co\{N(CH_2COO)_3\}_3]$
    Tris(nitrilodiacetate acetato-N,O)cobaltic(III) acid

$Na[Co(C_{10}H_{12}N_2O_8)]$
    Sodium (ethylenediaminetetracetato-N,N',O,O',O'',O''')cobaltate(III)

The designation of the number of oxygen atoms coordinated to the metal serves to indicate the structure. The question in the first case is whether the compound is 4-coordinate with two of the COO^- groups involved in coordination or 6-coordinate with four of the COO^- groups coordinated. The different names indicate these differences. The other examples require no further explanation.

The carbonyls and their derivatives are undoubtedly coordination compounds. However, their nomenclature has never followed that for coordination compounds. The following examples will show the ease with which this can be accomplished:

1	$Ni(CO)_4$	Tetracarbonylnickel(0)
2	$Co_4(CO)_{12}$	Duodecacarbonyltetracobalt(0)
3	$K[Co(CO)_4]$	Potassium tetracarbonylcobaltate($-I$)
4	$Co(CO)_3COH$	Tricarbonylhydrocarbonylcobalt($-I$)
5	$Fe(CO)_2(COH)_2$	Dicarbonyldihydrocarbonyliron($-II$)
6	$Os(CO)_4Br_2$	Tetracarbonyldibromoösmium(II)
7	$Re(CO)_3(py)_2$	Tricarbonyldipyridinerhenium(0)
8	$Re(CO)_3(py)_2Cl$	Tricarbonyldipyridinechlororhenium(I)
9	$[Co(C_5H_{11}N)(CO)_4]^+, [Co(CO)_4]^-$	Tetracarbonylpiperidinecobalt(I) tetracarbonylcobaltate($-I$) (27)
10	$Fe(CO)_3(SbCl_3)_2$	Tricarbonylbis(antimony trichloride)iron(0) (52)
11	$Ni(PCl_3)_4$	Tetrakis(phosphorus trichloride)nickel(0) (18)

The first two names are only inversions of the names presently used.

The third is a salt of a coordination entity in which the coordination center appears in an oxidation state of minus one.

Studies on the structures of the carbonyl hydrides indicate that the CO groups are essentially different from the —COH group; it seems proper to designate the two in different ways. The names suggested for compounds 4 and 5 indicate this.

Names 6, 7, and 8 are concerned with halogen derivatives and compounds in which CO has been displaced by other donor molecules.

Example 9 indicates the case of a compound with a coordination cation.

At such a time as the structures of the polymetallic carbonyls are known, the bridging CO groups can be designated by the Greek letter μ in the customary manner. It should even be possible to go further and indicate in the name that for $Fe_2(CO)_9$ (whose structure is known) there are three bridging CO groups and an Fe—Fe bond: hexacarbonyl-μ-tricarbonylbiiron (see Rule 14).

Of the various types of higher order compounds which should be named as coordination compounds, only molecular addition compounds have not been considered. The rules for naming inorganic compounds indicate that $(CH_3)_3N \cdot BF_3$ would be called "compound of boron trifluoride with trimethylamine." However, Davidson and Brown (4) suggest that this substance be called trimethylamine–boron trifluoride. In general, that molecule which donates a pair of electrons (base) is given first followed by the acceptor molecule (acid). Similarly, $(CH_3)_2S \cdot Al(CH_3)_3$ would be known as dimethyl sulfide–trimethylaluminum. If these were named as coordination compounds, the names would be: (trimethylamine)trifluoroboron and (dimethylsulfide)trimethylaluminum.

Compounds Not Readily Named as Coordination Compounds

There are some substances closely related to those discussed to which the coordination nomenclature pattern does not seem to be applicable. Some of these are excluded because of complexity, others because they do not follow the fundamental pattern of combination of coordination compounds.

There is no convenient way to name the metal derivatives of phthalocyanine, hemin, the chlorophylls, and similar compounds except to use special names for the two-dimensional super-ring systems which embed a metal ion at their centers. Among noncarbon compounds, super-ring systems (this time three-dimensional) are encountered among the heteropoly acids. Although the present system of naming these compounds gives no indication of structure, it seems hopeless to do this by any practical system of nomenclature. However, it is possible to indicate the central atom by calling them, for example, molybdophosphates instead of phosphomolybdates.

Those combinations which arise from lattice considerations alone do not follow the pattern of combination of coordination compounds. Hence, they are excluded from consideration here. These have been variously designated as lattice, clathrate (35), or occlusion (39) compounds, or as adducts (54).

Acknowledgment

The author wishes to acknowledge his indebtedness to his colleagues and graduate students who have offered suggestions and provided stimulating discussion during the formulation of the ideas set forth in this paper. He is particularly grateful to the friends and fellow scientists who read the preliminary manuscript and offered helpful comments. While it has not been possible to adopt all suggestions, the author hopes his choices were governed not by prejudice but by considerations of good nomenclature. His primary concern is not that of championing any particular set of rules but rather that of seeking general acceptance of some set which will be consistent with good nomenclature practice. The author also wishes to express his appreciation to the Research Corporation, Office of Naval Research, and the Atomic Energy Commission for the financial support of the experimental work which has kept these nomenclature problems ever before our laboratory. Some of the names which have been used for organic compounds are not those commonly used by inorganic chemists. The reason for this discrepancy arises from the author's strong feeling that a compound should be designated by the same name among both inorganic and organic chemists. In all cases, he has sought to follow the practice of *Chemical Abstracts*.

Literature Cited

(1) Bjerrum, J., "Metal Ammine Formation in Aqueous Solution," p. 80, Copenhagen, P. Haase and Son, 1941.
(2) Bragg, W. H., and Morgan, G. T., *Proc. Roy. Soc. (London)*, **104A**, 437 (1923).
(3) Cheesman, G. H., "Minutes of Meetings of Inorganic Nomenclature Commission," IUPAC Conference, 1949.
(4) Davidson, N., and Brown, H. C., *J. Am. Chem. Soc.*, **64**, 317 (1942).

(5) Emeléus, H. J., and Anderson, J. S., "Modern Aspects of Inorganic Chemistry," pp. 92, 95, New York, D. Van Nostrand Co., 1938.
(6) Ephraim, F., "Inorganic Chemistry," 3rd English ed. by P. C. L. Thorne and A. M. Ward, p. 289, London, Gurney and Jackson, 1939.
(7) Ewens, R. V. G., and Bassett, H., *Chemistry and Industry*, **1949**, 131-9.
(8) Fernelius, W. C., Larsen, E. M., Marchi, L. E., and Rollinson, C. L., *Chem. Eng. News*, **26**, 520-3 (1948).
(9) Finholt, A. E., Bond, A. C., Jr., and Schlesinger, H. I., *J. Am. Chem. Soc.*, **69**, 1199-203 (1947).
(10) Franklin, E. C., *Am. Chem. J.*, **47**, 285-317 (1912).
(11) Franklin, E. C., *J. Am. Chem. Soc.*, **46**, 2137-51 (1924).
(12) Franklin, E. C., "The Nitrogen System of Compounds," New York, Reinhold Publishing Corp., 1935.
(13) Gleu, K., and Brenel, W., *Z. anorg. u. allgem. Chem.*, **237**, 197-208, 326-34 (1938).
(14) "Gmelins Handbuch der anorganischen Chemie, No. 58B, Die Ammine des Kobalts," Berlin, Verlag Chemie G.m.b.H., 1930.
(15) Hein, F., "Chemische Koordinationslehre," pp. 33-5, Zurich, S. Herzel Verlag, 1950.
(16) Hückel, W., "Anorganische Strukturechemie," Stuttgart, F. Enke Verlag, 1948.
(17) Hückel, W., "Structural Chemistry of Inorganic Compounds," Vol. I, tr. by L. H. Long, Amsterdam, Elsevier Publishing Co., 1950.
(18) Irvine, J. W., Jr., and Wilkinson, G., *Science*, **113**, 742-3 (1951).
(19) Jaeger, F. M., "Optical Activity and High Temperature Measurements," New York, McGraw-Hill Book Co., 1930.
(20) Jorissen,, W. P., Bassett, H., Damiens, A., Fichter, F., and Remy, H., *Ber.*, **73A**, 53-70 (1940); *J. Chem. Soc.*, **1940**, 1404-15; *J. Am. Chem. Soc.*, **63**, 889-97 (1941).
(21) Ley, H., *Z. Elektrochem.*, **10**, 954 (1904).
(22) Mellor, J. W., "Comprehensive Treatise on Inorganic and Theoretical Chemistry," Vol. XIV, London, Longmans, Green and Co., 1935.
(23) Morgan, G. T., and Astbury, W. T., *Proc. Roy. Soc. (London)*, **112A**, 441 (1926).
(24) Morgan, G. T., and Drew, H. D. K., *J. Chem. Soc.*, **117**, 1457 (1920).
(25) Morgan, G. T., and Smith, J. D. M., *Ibid.*, **119**, 704 (1921).
(26) Ohmann, O., *Z. angew. Chem.*, **33**, I, 326-7 (1920); *Z. physik. chem. Unterricht*, **33**, 41-6 (1920).
(27) Orchin, M., private communication.
(28) Patterson, A. M., *J. Am. Chem. Soc.*, **55**, 3905-25 (1933).
(29) Pauling, L., "Nature of the Chemical Bond," 2nd ed., p. 7, Ithaca, Cornell University Press, 1942.
(30) Pauling, L., and Sherman, J., *Proc. Natl. Acad. Sci. U. S.*, **20**, 340 (1934).
(31) Pfeiffer, P., *Angew. Chem.*, **53**, 93-8 (1940).
(32) Pfeiffer, P., "Complexverbindungen," in K. Freudenberg's "Stereochemie," pp. 1200-377 Leipzig and Vienna, Franz Deuticke, 1932.
(33) *Ibid.*, p. 1356.
(34) Phillips, G. M., Hunter, J. S., and Sutton, L. E., *J. Chem. Soc.*, **1945**, 146.
(35) Powell, H. M., *Ibid.*, **1948**, 61-73.
(36) Remick, A. E., "Electronic Interpretations of Organic Chemistry," 2nd ed., p. 20, New York, John Wiley & Sons, 1949.
(37) Remsen, I., *Am. Chem. J.*, **11**, 298-306 (1889).
(38) Rosenheim, A., *Z. angew. Chem.*, **33**, Aufsatzteil 78-9 (1920).
(39) Schlenk, W., Jr., *Fortschr. chem. Forsch.*, **2**, 92-145 (1951).
(40) Schlesinger, H. I., and Shaeffer, R., "Report to the Office of Naval Research," 1951.
(41) Schwarz, R., "Chemistry of the Inorganic Complex Compounds," tr. by L. W. Bass, pp. 24-6, New York, John Wiley & Sons, 1923.
(42) Sidgwick, N. V., "Electronic Theory of Valence," pp. 152-62, London, Oxford University Press, 1929.
(43) Stock, A., *Z. angew. Chem.*, **32**, I, 373-4 (1919); *Ibid.*, **33**, Aufsatzteil 78-9 (1920).
(44) Sutherland, M. M. J., "Metal-Ammines," Vol. X of J. N. Friend's "Textbook of Inorganic Chemistry," London, Charles Griffin and Co., 1928.
(45) Taube, H., *Chem. Revs.*, **50**, 85-7 (1952).
(46) Vogel, A. I., *J. Chem. Soc.*, **1948**, 1833-5.
(47) Weinland, R. F., "Einfuhrung in die Chemie der Komplexverbindungen," 2nd ed., Stuttgart, Enke, 1924.
(48) Wells, A. F., *J. Chem. Soc.*, **1949**, 55-67.
(49) Werner, A., "Neuere Anschauungen auf dem Gebiete der anorganischen Chemie," 3rd ed., Braunschweig, F. Vieweg und Sohn, 1913.
(50) *Ibid.*, pp. 92-5.
(51) Wheland, G. W., "Advanced Organic Chemistry," 2nd ed., p. 224, New York, John Wiley & Sons, 1949.
(52) Wilkinson, G., private communication.
(53) Wittig, G.,"Stereochemie," pp. 226-75, Leipzig, Akademische Verlagsgesellschaft m.b..H, 1930.
(54) Zimmerschied, W. J., Dinerstein, R. A., Weitkamp, A. W., and Marshner, R. F., *Ind. Eng. Chem.*, **42**, 1300-6 (1950).

RECEIVED November 3, 1951.

Problems of an International Chemical Nomenclature

K. A. JENSEN
Chemical Laboratory, University of Copenhagen, Copenhagen, Denmark

Is it possible to obtain a much more radical unification of chemical terms than has hitherto been attempted? This article discusses approaches to the problems of international agreement on spelling, pronunciation, the abandoning of national names, and the naming of elements and salts. Only on etymological spelling can there be agreement. The problems of transliteration of Greek words require the formulation of orderly rules. The substitution of new chemical names formed from Greek or Latin roots for national names would be a significant step toward standardization. The adoption of the Stock nomenclature system with Latin names for the elements and the use of consistent terminations for organic compounds are suggested. A realistic approach to the unification problem, however, allows for logical deviations from general rules, when the exceptions are agreed upon.

Interest in chemical nomenclature has been increasing in recent years. The marked activity on all fronts of chemistry has increased the demand for effective communication, for clearness and precision in information, and has thus stimulated interest in chemical nomenclature problems. International cooperation on chemical problems has at the same time emphasized the need for more conformity in the chemical nomenclature of different languages.

The Geneva conference of 1892 initiated chemical nomenclature work on an international scale, and this was later continued under the auspices of the International Union of Chemistry but was interrupted and severely hampered by two world wars and their depressing influence on international cooperation and confidence. Nevertheless the International Union of Chemistry succeeded in accomplishing nomenclature work which has already exercised a far-reaching influence on chemical literature.

On several points, however, the committees of the International Union of Chemistry have refrained from attempts to create a truly international chemical nomenclature, leaving it to national committees to modify the adopted names in conformity with "the genius of each language." In doing this, problems of a purely linguistic nature have been highly overestimated. Actually there is, from a modern philological point of view, no fundamental reason for not introducing a much more radical unification of chemical terms, at least among the Indo-European languages. Among these languages there is a unanimity of major pattern which in principle makes it possible to formulate an idea in much the same way. There is no real reason why the French should not say "ferrochloride" instead of *chlorure ferreux*, the Italian "ferro-clorido" instead of *cloruro ferroso*, the Czech "ferro chlorid" instead of *chlorid železnatý*, etc. Only habit prevents the use of essentially the same name in all these languages. Similarly, no rule of language forbids the Italians to write "helio" and "oxido" instead of *elio* and *ossido*, the Spanish to write "scandio" and "yttrio" instead of *escandio* and *itrio*, and the Germans to write "Citronensäure" instead of *Zitronensäure*. There is no more reason for some languages to continue to use names such as benzol, kolesterin, and dioxyacetone than there was to use irrational names such as carbolic acid and spirit of wine.

Abandonment of National Names

It is now commonly agreed that a scientific nomenclature should be as international as possible. Formerly, however, it was thought to be a point of national self-assertion to introduce national terms. In some languages (Finnish, Icelandic, Czech, etc.) almost all foreign words have been replaced by purely national words—even such words as the names of the months, electricity, cigar—which otherwise pass almost unchanged from language to language. In many languages, the names hydrogen, nitrogen, oxygen, and carbon have been translated (German: *Wasserstoff*, etc.) or replaced by newly created national words—Danish, *brint* (hydrogen), *ilt* (oxygen), and Polish, *tlen* (oxygen). The Slavic languages also have national names for silicon, aluminum, magnesium, and calcium. This again gives national names to the inorganic compounds. From Danish *ilt* (oxygen) is derived *ilte* (oxide). Thus names such as *brintoverilte* and *kvælstofilte* are obtained for compounds for which the names "hydrogenperoxid" and "nitrogenoxid" could be used just as well. This has now actually been proposed by the Danish nomenclature committee. Czech analogously has a special name for oxide, *kysličník*, and thus names such as *kysličník vápenatý* (calcium oxide) and *kysličník uhličitý* (carbon dioxide) are obtained.

Unification of the names of the chemical elements would, in principle, do away with all these inconsistencies.

In older times, most chemical terms were taken from words of daily life and thus differed from language to language—"slaked lime," "sugar of lead," "butter of antimony," "heavy spar," "massicot," etc. On the introduction of rational terms, the old names have almost disappeared from scientific literature. In several languages, however, names of some organic compounds, particularly organic acids, still are reminiscent of the names of early chemistry. For succinic acid, for instance, names such as *Bernsteinsäure* (German), *ravsyre* (Danish), and *yantarnaya kislota* (Russian) are used. In English and French, Latinized names of these acids are used, and these could as well be applied in other languages—e.g., "Succinsäure" instead of *Bernsteinsäure*. Even if it is difficult now to change these old names, new names should at least not be formed in the old-fashioned way. When the name folic acid had to be used in other languages, some authors translated it into *Blättersäure* (German) and *bladsyre* (Danish, Norwegian). Rapidly, however, the names *Folinsäure* and *folinsyre* gained ground.

It would be very important if the International Union of Chemistry would state as a principle that new chemical names should always be formed from Greek or Latin roots. When introducing new names, due consideration should be given to their applicability in other languages.

These rules should also apply to chemical terms that are not names of compounds. Terms such as "lone pair," "shared electrons," "core," *Zwitterion, Eigenfunktion, Bindigkeit, Ionen-Beziehung*, and *Oktett-Lücke* are particularly difficult to translate and often are taken over unchanged in other languages, a practice which can only be considered as makeshift. *Zwitterion* is rendered in Danish as *ampho ion*, a word of real international form which should be introduced in other languages.

For most of the organic functional groups, terms of international character are used. This, however, is not the case with the name "acid," for which very different names are used—*Säure, syre, happo, kwas, kislota*, and *kyselina*. It would be preferable to introduce a term of Latin origin—as has actually been done in the Turkish *asid*—but in several languages it is not possible to distinguish in pronunciation between "acid" and "azid."

Words of a national character should not be connected with international names. In Polish, for instance, the national numerals are used instead of mono, di, tri, etc., resulting in the names *dwunitrotoluen* and *czterohydropirano* for dinitrotoluene and tetrahydropyran. The inconvenience of this is obvious, but no less inconvenient is the use of national names such as iron, lead, *Wasserstoff*, and *Bernstein* in connection with international terms such as oxid(e), sulfat(e), and amid(e).

Names of the Elements

One of the most needed improvements in inorganic chemical nomenclature is the unification of the names of the chemical elements. If essentially the same names of the elements could be introduced in different languages, the names of inorganic compounds in these languages would automatically become very similar.

The differences in the names of the elements in different languages are of the following types:

1. The elements known from antiquity (gold, silver, copper, iron, mercury, lead, tin, and sulfur) have different names in different languages.
2. For some of the more recently discovered elements (hydrogen, carbon, nitrogen, and oxygen) several languages employ translations of the original names of international character.
3. In some cases, two names of international form have been introduced for the same element:

        Beryllium and glucinium   Lutecium and cassiopeium
        Sodium and natrium   Hafnium and celtium
        Potassium and kalium   Antimony and stibium
        Columbium and niobium   Tungsten and wolfram

4. In several languages, the spelling of the names of the elements has been changed by applying a more or less phonetical spelling. As the pronunciation of the names of the elements differs from language to language and the phonetical value of the letters is not the same in different languages, these changes have made the names look different, not only from the Latinized names, but also from each other:

k replaces c: kalcium, kadmium, skandium, aktinium, kobalt (Scandinavian, German Slavic languages).

k replaces ch: klor, krom (Scandinavian, Croatian, Hungarian, Turkish).

z replaces c or vice versa: Kalzium, Zäsium (German); cynk, cyrkon (Polish).

s replaces c or z: kalsium, sink (Norwegian, Finnish); seziyom (often in Turkish instead of cesium).

h is eliminated: litium, renium, tallium, torium, tulium, rodium, rutenium, lantan (several languages); Italian elio (He) and afnio (Hf).

i replaces y or vice versa: berillio, kripto, itrio, iterbio (Italian and Spanish); tytan, iryd, cyrkon, cynk (Polish).

The ending -ium is altered: into -io (Italian, Spanish); -iu (Rumanian); -iĭ (Russian); or completely omitted: Niob, Zirkon (German); nioob (Dutch); lit, skand, gal, osm, ren, rad (Polish).

To obtain practically the same names in different languages, the following changes are necessary:

Introduction of Latin Names. The Latin names of the elements of the first category (*aurum, argentum, cuprum, ferrum, hydrargyrum, plumbum, stannum,* and *sulfur*) should be introduced as scientific names into the literature of all countries. These should not necessarily replace the national names in common usage. The use of Latin names may be of importance in distinguishing between the element ferrum and common iron, for example, but it will be an advantage to use these names especially when forming names of compounds.

In English, the Latin names sulfur and mercury have practically superseded the names brimstone and quicksilver, even in common usage. Unfortunately, however, the name mercury is at variance with the chemical symbol.

Replacing of National Names. The names of this group are artificial names that have not become part of everyday language to the same extent as the names of the first group. It should be possible to replace national names (*Wasserstoff*, etc.) by the original names hydrogen, oxygen, nitrogen, and carbon.

Selection of a Single Form. The problems concerning the names of this group are in some cases mingled with a conflict of priority, and therefore involve national prestige. In many cases, however, the question of priority cannot be settled unambiguously. A chemical name should not be accepted on grounds of priority alone; its adequacy, applicability, and present usage in different languages should also be considered.

At the Amsterdam meeting of the IUC (September 1949) it was decided to recommend the names beryllium, niobium, lutetium, hafnium, and wolfram. The cases where there are two different names would thus be reduced to three. Unfortunately, opposition to the name wolfram, especially from the English side, caused the committee to leave this name optional again. The reasons for choosing these names are the following:

BERYLLIUM. The name glucinium has been little used outside France and is not particularly characteristic for this element. The name beryllium was proposed by Wöhler, who first prepared metallic beryllium. The priority of Vauquelin for being the first to prove a new "earth" in beryl is not contested by this decision.

NIOBIUM. The name columbium has only been used by American and some British authors. The name niobium, proposed by H. Rose who was the first to give definite proof of the existence of this element, is used in all other languages. On grounds of priority, there is no reason for preferring the name columbium (Larsson, *3*).

LUTETIUM. The name cassiopeium is used only in German and (in part) Dutch. Both for this reason and because the name is difficult to adapt to some languages, the name lutetium is preferred. It is not intended to settle the question of priority (Paneth, *5*) by this decision. The hitherto used spelling "lutecium" should be replaced by lutetium, because the name should be derived from the Latin form *Lutetia* rather than from its French equivalent *Lutèce*.

HAFNIUM. The name hafnium is used in all languages other than French. It is also preferred on grounds of priority.

WOLFRAM. Wolfram is the name proposed by the discoverers of this element, the brothers de Elhuyar. It is in accordance with the chemical symbol and is used in 12 or 13 languages out of 17 considered. On the basis of his historical studies, Moles (*4*) has strongly advocated the general introduction of the name wolfram, and even in England this name has been recommended (Hadfield, *2*). Further, tungsten in Scandinavian languages means "heavy stone" and could not well be applied to an element, since it is the original and still used name for the wolfram mineral scheelite, $CaWO_4$. Thus the arguments for abandoning the name tungsten for the element are very strong.

In the remaining cases where there are two names for the same element, consideration of the accepted symbols favors the names natrium, kalium, and stibium instead of sodium, potassium, and antimony.

Regulation of Spelling. Spelling should be regulated. In the main languages, only minor changes in spelling are necessary:

The spelling "baryum" is used only in French and Czech, where it should be changed to barium. In English, sulfur should not be spelled with ph. In American, aluminum should be replaced by "aluminium."

In recent German, there has been a tendency to use k and z instead of c (*Kalzium, Zäsium,* and *Kadmium*). Since the war, however, the Latin forms are gaining ground again. "Vanadin" should be changed to Vanadium, "Niob" to Niobium, and "Wismut" to Bismuth.

In other languages, more extensive changes in spelling are necessary to arrive at true international names.

The introduction of these names would mean a major step toward an international chemical nomenclature. Thus "sulfurtrioxid(e)" would replace *Schwefeltrioxyd, zwaveltrioxyd, svovltrioxyd, kyslíčnik sirový,* and *trekh-okis sery.* "Hydrogenperoxid(e)" would replace *Wasserstoffperoxyd, brintoverilte, vätesuperoxid,* and *perekis vodoroda.* "Nitrogenoxid(e)" would replace *Stickstoffoxyd, kväveoxid, oxyde d'azote, kvælstofilte, kyslíčnik dusnatý,* etc.

The Stock Nomenclature

According to the IUC rules for naming inorganic compounds:

Indication of the electrochemical valency in the names of compounds should be made only by Stock's method and the system of valency indication by terminations such as -ous, -ic, etc., should now be avoided not only in scientific but also in technical writing.

This report was published in 1940, when work in the International Union was suspended because of the war, and could not therefore be thoroughly discussed and criticized before an international forum. In the following years, the Stock system became widespread, especially in Germany and the United States.

From an international point of view, however, the Stock nomenclature suffers from the serious drawback that it is more national in character than the older nomenclature. This may be substantiated by giving the names of $FeSO_4$ in different languages according to the older nomenclature and the Stock system:

Language	Older Name	Stock Name
Danish	Ferrosulfat	Jern-to-sulfat
Dutch	Ferrosulfat	Ijzer-twee-sulfaat
English	Ferrous sulfate	Iron-two-sulfate
Finnish	Ferrosulfaatti	Rauta-kaksi-sulfaatti
French	Sulfate ferreux	Sulfate de fer-deux
German	Ferrosulfat	Eisen-zwei-sulfat
Hungarian	Ferroszulfát	Vas-kettö-szulfát
Italian	Solfato ferroso	Solfato di ferro-due
Russian	Sulfat zheleza	Sernokisloe zhelezo-dva
Spanish	Sulfato ferroso	Sulfato de hierro-dos
Swedish	Ferrosulfat	Järn-två-sulfat
Turkish	Ferrosulfat	Demir-ikki-sulfat

Accordingly, the failure of the Stock system is that it uses the national names of the elements and, in spoken language, the national numerals. In written language, the numerals will be designated by Roman figures, but even then the Stock nomenclature is much more national than the one hitherto used. The joint Scandinavian (Danish, Finnish, Norwegian, and Swedish) nomenclature committee wrote on this matter to the International Union of Chemistry:

We find this objection so significant that we question the advisability of recommending the Stock nomenclature without radical alteration of it. At any rate we find it objectionable that the international rules mention Stock's system as the only one allowed for indicating the electrochemical valency.

In the written language, the designation by Stock's method can be made international by the use of the Latin names of the elements—e.g., ferrum(II)-sulfate, cuprum(I)-chloride, stannum(IV)-chloride, and aurum(III)-chloride. Contrary to the English version of the international rules, it is considered advisable to use a hyphen after the parenthesis.

Fernelius (1) says that there is a definite tendency to shift to English names throughout—for example, to use silver(I) ion instead of argentous ion and iron(II) ion instead of ferrous ion. This is certainly a step backward. According to Fernelius, "The percentage of such names is so small that any benefit a foreign reader may be expected to gain is negligible." In a chemical text, however, the chemical terms are so important that often little more than the understanding of the chemical terms is necessary to get the meaning of a chemical text in a foreign language. Furthermore, it is an advantage to the chemists of the small countries who are compelled to write in a language other than their own that the chemical terms look similar in different languages.

The proposal to use the Latin names of the elements seems to be the only satisfactory solution of this intricate matter. It is true that it does not do away with the difficulty that national numerals have to be used, but as chemical knowledge is mainly communicated in written language, this objection is of minor importance.

The names ferrum(II)-chloride, etc., resemble very much previously used names, and indeed the endings -ous and -ic may be used when they leave no uncertainty as to the valence. The circumstance that the ending -ous sometimes refers to a univalent, sometimes to a bivalent, and sometimes to a trivalent state is no serious objection. Often it is necessary to indicate more than two valencies, and there the ous-ic system fails. Werner (*6*) proposed an alteration and extension of the o-i- system, using the endings a-o-i-e-an-on-in-en, thereby indicating 8 different valencies. This ingenious system failed because the vowels in some languages, particularly English, are not pronounced distinctly enough. In Czechoslovakia, a system is in use in which the name of the metal is used in adjective form with different endings, indicating different valencies. Thus the oxides MnO, Mn_2O_3, MnO_2, MnO_3, and Mn_2O_7 in Czech are called *kysličnik manganatý, manganitý, manganičitý, manganový,* and *manganistý*. This nomenclature represents a purely national solution of the problem and indeed is an example of the way in which a scientific nomenclature should not be developed.

Thus it seems to be most satisfactory to use the Latin names of the elements with the Stock numbering system.

Names of Salts

In the Stock nomenclature, the acid residue is indicated in the same way as in the o-i- system. The terminations -at(-e,-o) and -it(-e,-o) are now universally adopted, and the names formed in this way have almost completely superseded the names of the types "schwefelsaures Magnesium" or "schwefelsaures Magnesia," formerly used in the German, Russian, and Scandinavian languages. The termination -id(-e,-o), however, is not universally adopted. The Romance languages (and Turkish) use instead the termination -ur (-e,-o). This was formerly also used in English and German, where it, however, was used to designate a lower oxidation state (*Eisenchlorür*, $FeCl_2$; *Eisenchlorid*, $FeCl_3$).

When by the introduction of the o-i- system the valency state was indicated by terminations of the name of the metal, one of the terminations -ur and -id became superfluous, and in most languages the termination -id(-e,-o) was chosen. Actually, it is difficult to distinguish in pronunciation between -id(-e,-o) and -it(-e,-o). When the d is not followed by a vowel, it is generally pronounced as t and recourse has to be taken to modify the length of the i to distinguish between -id and -it. In some languages, however, a certain vowel can have only one length, and in these languages it is absolutely impossible to distinguish between "chlorid" and "chlorit," "sulfid" and "sulfit". Even when the d is followed by a vowel (as in English) the difference between -ide(-o) and -ite(-o) is not very distinct.

Therefore, the reintroduction of the termination -ur(-e,-o) would offer certain advantages. Unfortunately, however, the termination -id(-e,-o) is much more widespread and even in the Romance languages the name "oxyde(-o)," not "oxure(-o)," is used.

Perhaps the best solution of this problem would be to choose a new termination. By analogy with the name *oxyd(e)*, which is used in French, German, and some other languages, "chloryde," "sulfyde," etc., might be written. In the languages where this difficulty of distinguishing arises, the termination -yd could be pronounced, and perhaps written, as -üd.

It would be an advantage if all compounds of this type were designated in uniform manner in all languages. The difficulty in attaining this is unusually severe, since the changes will be great, whatever decision is made:

If the termination -id(-e,-o) is retained, the termination -ur(-e,-o) in Romance languages should be changed and *oxyd(e)* should be changed into "oxid(e)." The problem of distinguishing clearly between -id and -it is not solved in this manner.

The introduction of the termination -ur(-e,-o) means a change in all languages other than the Romance ones, and also for these "oxur(-e,-o)" is a new name.

The introduction of the termination -yd(-e,-o) means a change in all languages, but this termination is not very different from -id(-e,-o) and may be pronounced as this. In the languages where a different pronunciation is desirable, it may be pronounced as -üd.

The names indicating the anion can be combined with the name of the metal in two ways, illustrated by magnesium sulfate and sulfate of magnesium. In French, exclusively, names of the latter type are used (*sulfate de magnesium*). The names of the first type are most convenient when the acid has a short name, and they could well be employed in French also. When, however, the acid has a very long name, the distance between the name of the metal and the ending -ate becomes unduly long, and in such cases it seems more convenient to use names of the French type, "3,5-dinitro-4-dimethylaminobenzoate of sodium" (in German: "Natriumsalz der 3,5-Dinitro-4-dimethylaminobenzoesäure" or "3,5-dinitro-4-dimethylaminobenzoesaures Natrium").

The same applies to esters where names of the type benzoic ethyl ester (in German: "Benzoesäureäthylester") are to be preferred to names of the type ethyl benzoate, when the acid has a long name.

Terminations of Names of Organic Compounds

Building on the Geneva rules, the International Union of Chemistry in its "Definitive Report of the Commission on the Reform of the Nomenclature of Organic Chemistry" (the "Liége Report") has created a consistent system of terminations expressing the functions of organic compounds. This system was first adopted in English literature and was also introduced in French and (in part) in Dutch, Spanish, and Polish. It is also being introduced (as far as possible) in the Scandinavian languages. Although the Geneva rules were created also on the initiative of German chemists, German literature continues to use the names benzol, toluol, carotin, cholesterin, glycerin, resorcin, etc. It is true that the Liége rules cannot be applied so consistently to the German language as to English but the names mentioned could be altered as proposed in the Geneva rules.

The difficulty of a consistent application of the Liége rules is connected with the termination -e, which in German indicates the plural. Therefore, it is not possible to use the endings -an and -ane, -in and -ine, -ol and -ole as designations of different functions. *Oxazole* in German is simply the plural form of *Oxazol*. Similar difficulties, however, are encountered in Scandinavian and other languages, where it has been found advantageous to introduce the system as far as possible.

One of the most urgent problems for the creation of an international chemical nomenclature is the universal adoption of the endings -en(-e,-o) for unsaturated hydrocarbons ("benzen," "toluen," "styren," "caroten," etc.) and the ending -ol(-o) for hydroxy compounds (glycerol, resorcinol, cholesterol, etc.) On the other hand, it is not possible in several languages to avoid giving heterocyclic compounds the same ending as hydroxy compounds (pyrrol-glycerol) or saturated hydrocarbons (pyran-methan) or to give some nonnitrogenous compounds the same ending as nitrogen bases (dextrin-pyridin). Even with this limitation the Liége rules offer such advantages that it is unwise to neglect them.

The prefix hydroxy- should be universally adopted to express the alcohol or phenol function. In German and several other languages (presumably under the influence of the Beilstein nomenclature) the prefix oxy- is used (*Oxysäuren* = hydroxy acids, *Oxybenzole* = hydroxybenzenes). This is a relic of the old addition nomenclature, oxy- meaning an oxygen atom (trioxymethylene, phosphorus oxychloride), but organic chemical nomenclature now generally follows the principle of substitution.

The prefix hydroxy-(oxy-) should not be spelled "hydroxi-(oxi-)."

Spelling of Chemical Terms

Agreement on Spelling of Beginning of Words. The international committees of chemical nomenclature have concentrated their efforts on attaining uniformity in the endings of chemical names. In this respect, they have been successful. The beginnings of the names have been left as a problem of orthography which cannot be dealt with by chemists, at least on an international level. For the perception of the names, however, the beginnings of the words are as significant as the endings, simply because words are read from left to right. There is certainly as much difference between "cholesterol" and "kolesterol" as between "cholesterol" and "cholesterin," between "caroten" and "karoten" as between "carotene" and "carotin," between "phloroglucinol" and "floroglucinol" as between "phloroglucinol" and "phloroglucin."

It is particularly inconvenient, both as to recognizability and to mnemonics, when chemical names of plant substances are spelled in another way than the botanical names from which they are derived. In some languages the names of compounds isolated from Colchicum, Crocus, Coca, Citrus, Rhamnus, Rheum, Quercus, Physalis, Thymus, Haematoxylon, Hedera, and Chelidonium are spelled *Kolkicin, krozetin, kokain, zitral* or *sitral, ramnose, rein, kversetin, fysalien, tymol, ematossilina, ederagenina,* and *kelidonin.* For chymosine, *chimosina, quimosina,* and *kymosin,* for chlorophyll, *clorofilla* and *klorofyl* are found.

Conformity in spelling results in the same alphabetical order of chemical terms, irrespective of the language. A chemist frequently uses compilations of chemical data and subject indexes of journals and books of reference in languages other than his own. It is inconvenient to search for "quinoline" under q in an English journal, but under ch (*Chinolin*) in a German and under k (*kinolin*) in a Scandinavian journal. Even such small differences as "thorium" and "torium," "rhodium" and "rodium," "rhamnose" and "ramnose" are inconvenient and may cause important references to be missed.

Finally, chemists of small countries, who are obliged to write papers in a main language, are hampered by national spelling. Scandinavian chemical literature, written in English, sometimes contains such misspellings as "bensoic" acid, "bensyl," "styfninic" acid, "tiazole," "metyl," etc. Influenced by the chemical symbols and by papers written in foreign languages Scandinavian students of chemistry are inclined to use spellings such as "calcium," "chlor," "chrom," and "phenthiazin," but these have been corrected to "kalcium," "klor," "krom," and "fentiazin." When these spellings have been impressed on the students, they add to the confusion when the students have to read or write in a foreign language.

From the point of view of mental effort, the unification of the names of the chemical compounds offers great advantages. Because the chemical symbols are international, efforts should primarily aim at introducing names of the elements which are in accordance with their symbols.

Etymologic versus Phonetic Spelling. Divergent spellings are particularly numerous in the languages of small European countries where the written language has no old tradition, but has mainly been established in the 19th century. At that time, the predominant opinion of philologists was that a written language should, as far as possible, be a phonetic rendition of the spoken language. But as pronunciation differs from language to language, and the phonetical value of a certain letter is not the same in different languages (c is pronounced as s or k in western languages, but as ts in Slavic languages; z is pronounced as ts in German, but as voiced s in Slavic languages), the introduction of phonetic spelling has isolated these languages from each other and from the classical languages. The newer theory of language, however, considers the spoken and written language as two different coordinated modes of expression and does not approve of efforts to bring them as close together as possible.

The process of reading is not, as was formerly thought, simply a process of spelling. In reading, words are not spelled but are perceived as entities, as symbols. It is not absolutely necessary that these symbols have any connection at all with the sounds for which they stand—as is well known from Chinese ideographs and from chemical symbols or numeral figures. A certain approach to a phonetical rendition has some value as to mnemonics, but an exact phonetical rendition of the spoken language is impossible because the pronunciation of a certain word may vary with place, time, and even the connection in which it is used. Because words are read as symbols, it is of importance that the same thing is always described by the same symbol.

A written language has, as a rule, only one accepted spelling of each word even if there may be several pronunciations. The advantage of this is uncontested when a single language is concerned, but it also applies to international words occurring in different languages. The pronunciation of "lieutenant" is quite different in French and English, but it is obviously an advantage to understanding that the "symbol" is the same. Nevertheless, many languages have nationalized the spelling of such international words.

An alternative to the phonetical spelling of words is an etymological spelling, in accordance with the derivation from the parent language. As a matter of principle, scientific terms should as far as possible be derived directly from the classic languages and not from modern languages. In some instances, terms have been formed from words of modern languages and should retain the spelling of the language of origin (*ytterbium*, named after the Swedish town Ytterby).

Phonetic spelling will vary from language to language; the etymologic spelling may be constant. An etymological spelling is the only spelling on which there is the possibility of an international agreement. A difficulty, however, arises with terms derived from languages that are not written with the Latin alphabet: Greek, Arabic, Chinese, and Russian. The etymological spelling of words derived from these languages will be constant only if there is agreement concerning their transliteration. The foundation of the sciences was laid in a literature written in Latin and in this literature words of Greek (and Arabic) origin were transcribed by a consistent and universally accepted system. It is rational to continue to use this system, even if it does not give the simplest spelling possible in some cases.

English and French have to a large extent retained the classic spelling of words of Latin and Greek origin, even if the pronunciation has changed from the original one. German formerly had retained the classic spelling of words of Latin and Greek origin, but in more recent times the spelling has been nationalized by introducing z and k instead of c (*Krozetin*, *Kalzium*, *Zäsium*, *Kapronsäure*, *Karbazol*, *Karboxyl*, *Kozymase*, *Kumarin*, *Zellobiose*, *Zetylalkohol*, *Zitronensäure*, *Zitral*, *Zyan*, *Zyklohexan*, and *Zystein*). This spelling has also spread to a great extent in Swiss literature. The German handbooks, Beilstein, Gmelin, and Richter-Anschütz, however, use the older spelling and since World War II this spelling is gaining ground again, so there should be a fair chance of bringing the German spelling of chemical terms very close to the English and French.

The main difference between the English-French and older German spelling is the rendition of Greek kappa and the spelling of names related to quinine (quinoline, quinic acid, quinone, quinhydrone, quinuclidine, quinoxaline, inter alia), which in German are spelled with ch. The common origin of these words is a Peruvian word *kina* (bark), but as this has come into European languages via Spanish, where it is spelled *quina*, it seems most logical to use the spelling with qu and not that with ch.

Transliteration of Words of Greek Origin. The peculiarities of the different spellings of chemical terms are connected with a different transliteration of words of Greek origin. Thus zeta may be reproduced as z, s, or c; theta as th or t; kappa as c, k, s, or z; xi as x or ks; rho as r or rh; upsilon as y or i; phi as ph or f; and chi as ch or k. In words of Latin origin, c may be reproduced as k and qu as kv.

It would be of great importance if the International Chemical Union could recommend only one system to be used when forming chemical names from Greek roots. The most rational system would be to use the Latin transliteration (kappa = c, phi = ph, chi = ch, etc.).

German, in contrast to English and French, often reproduces kappa as k, when it is pronounced as k (*Pikrinsäure*, *Kresol*, *Kakodyl*, often also *Galaktose*, *Oktan*). It is difficult for the Greek kappa to be generally reproduced as k. In chemical names introduced in older times, the Latin transliteration of Greek words was used exclusively, and accordingly the pronunciation of kappa before front vowels (e,i,y) changed to s (ts)—for instance, cetyl, decyl, cinnabar, glycin, cyan, cyclo-, cystin, cytochrom. It does not seem advisable to change these spellings to "ketyl," "dekyl," "kinnabar," "glykin," "kyan," "kyklo-," "kystin," "kytokhrom." In the words where c is pronounced as k the spelling could in many cases be changed, as has been done in German and several other languages, but in some words the c has been stabilized by becoming part of a chemical symbol. It is unwise to introduce "Kadmium" and "Aktinium" as long as their chemical symbols are Cd and Ac.

In the international, Latinized names of zoological and botanical species the Latin transliterations for Greek names are still used. When the name of a chemical compound is derived from these names, it is important that the spelling should not be altered. For instance, "crocetin" should be spelled thus, because its name is derived from the botanical name Crocus, although this again is derived from Greek *krokos*. The same argument applies to colchicine, coniine, corydaline, and many others.

It is not possible to reproduce kappa as k consistently. The most rational procedure is to use the classic Latin transliteration of Greek words—i.e., practically not to use the letter k in chemical names. This is in accordance with French, other Romance languages, and, for the most part, with English. In modern English chemical nomenclature there is a tendency to reproduce kappa as k—for instance, katalase, hexokinase, kathepsin, keratin, kynurenine, kyanmethine, kephalin. To avoid further confusion it is very important that a rule should be established concerning the formation of chemical names from Greek roots.

There is some merit in the simplified reproduction of phi as f. However, ph will presumably be used forever in the names of botanical and zoological species and in medical terms, and it is desirable that chemical names derived from these terms should conserve the same spelling (physostigmine, physalin, phthiocol). It is unwise to spell phosphorus with f as long as the chemical symbol is P. It therefore seems most rational to adhere strictly to the Latin transliteration of Greek words and to spell words of Latin origin in the same way as the words from which they are derived.

A system of transliteration should be considered not as an end but as a means, and exceptions to it may therefore be admitted, if they are agreed to internationally. The spellings krypton, barium, ethyl, kynurenin are found in more languages than the more orthodox spellings "crypton," "baryum," "aethyl," "cynurenin," and it would therefore be easier to obtain international agreement concerning the first spellings. Such exceptions should, however, not be too numerous.

Languages with Phonetic Spelling. Next to English, French, and German the languages of greatest importance to chemistry are Russian, Italian, Spanish, Dutch, Polish, and Czech. Russian need not be considered here, because the use of another alphabet makes almost any approach to an international spelling impossible. In Dutch almost no changes are necessary. Polish has strictly phonetical spelling with k instead of c, ks instead of x (even when an initial sound: *ksenon*, *ksantogen*) and f instead of ph. A peculiarity of Polish is that i has been replaced by y and vice versa (pyridine becomes *pirydyna*). Thus rather extensive changes are necessary. Czech has a spelling of international words which is to a great extent in accordance with the international spelling, though for many compounds purely national names have been introduced.

In Italian the most serious deviations from the international spelling are the disuse of the letter h and the replacement of x by s or ss and of y by i. Thus such names as *esaossicicloesane* (hexahydroxycyclohexane) arise. H is not pronounced in Italian, but nevertheless is retained in some purely Italian words (*ho*, *ha*), so it should be possible to replace names such as *elio*, *afnio*, *emina*, *emoglobina*, *eptacosano* by "helio," "hafnio," "hemina," "hemoglobina," "heptacosano." By reintroducing x and y such names as *esaidropirano*, *esosi*, *ossiemoglobina* would be "hexahydropyrano," "hexosi," "oxyhemoglobina," and thus much more easily recognizable.

In Spanish the most serious deviations from international spelling are caused by the use of i instead of y, by insertion of an e before s followed by another consonant, and by sometimes reproducing ch as qu. Thus names as *itrio* (yttrium), *escandio*, *estroncio*, *escopolamina*, and *quimosina* (chymosine) arise.

In the languages having a phonetical spelling, the introduction of an etymological spelling is often rejected with the argument that it is not in accordance with the "genius of each language." This argument has, however, been grossly abused in discussions of chemical nomenclature. The Latinized botanical and zoological names are used unchanged in different languages. In Italian the botanical names Oxalis and Helenium are accepted unhesitatingly; why, then, not "oxigeno" and "helio"? In all languages proper names exist which do not conform with a phonetical spelling. The names of the chemical elements and compounds are in some sense to be considered as proper names.

To introduce the changes necessary to obtain an international spelling of chemical terms it is necessary to consider these as Latinized scientific terms similar to the names of botanical and zoological species. From this point of view, it should be possible to change the spelling of chemical terms without changing the spelling of words of everyday life of Greek or Latin origin, and conversely, when first established, the international spelling of chemical terms should not follow changes in the spelling of common words.

Pronunciation

A real unification of the pronunciation of chemical terms is not attainable, nor is it of any great importance because most chemical communication is given in written language. French and German are in some sense better suited as international languages because they are pronounced more distinctly. Without doubt, however, English is becoming more and more important as a congress language, and if this is to be for the good, English and American chemists should make efforts to articulate more distinctly.

The importance of English as an international language is also severely hampered by the fact that often international words which are pronounced in much the same way in several languages are pronounced differently in English. Names as amid(e,-o), methyl, and cyanid(-e) sound very similar in French, German, Italian, Russian, Danish, etc., but quite different in English. The cause of this is that i and y in English may be pronounced as ai, a pronunciation which is different from that of all other languages, and even from the original Anglo-Saxon one. From this point of view the American pronunciation of methyl, phenyl, etc., is much to be preferred to the English one. The AMERICAN CHEMICAL SOCIETY has issued a list of recommended pronunciations, and in many cases two possible pronunciations have been placed in order of preference. In almost all cases, from an international point of view, the order of preference should be reversed, because the specific English pronunciation has been placed first. Some pronunciations have been described as Germanisms. These pronunciations resemble not only the German pronunciation, but also that of other languages and therefore their use should be encouraged rather than discouraged. A term particularly difficult for foreigners to understand is thio-. It would be an advantage if it could be decided to pronounce it teeo- as in most other languages.

Influence of IUPAC

The difficulties encountered in the establishment of an international chemical nomenclature may appear insuperable to those who are strongly dependent on habit and to those who consider the language as a growing organism which cannot be influenced arbitrarily. The latter opinion is erroneous. Which word should be used for a certain notion is almost exclusively a question of convention, but it must be admitted that habit will strongly oppose all attempts at radical changes of language. Notwithstanding this difficulty, the nomenclature committees of the International Union of Pure and Applied Chemistry should seriously consider the possibilities of unification of chemical terms, or their influence will be limited to only a few languages.

Acknowledgment

The author acknowledges the assistance of V. E. Price and G. H. Cheesman.

Literature Cited

(1) Fernelius, W. C., Larsen, E. M., Marchi, L. E., and Rollinson, C. L., *Chem. Eng. News*, **26**, 521 (1948).
(2) Hadfield, R. A., *J. Iron Steel Inst. (London)*, **64**, 69 (1903).
(3) Larsson, A., *Z. anorg. Chem.*, **12**, 193 (1896).
(4) Moles, E., *Anales soc. españ. fís. y quím.*, **26**, 234 (1928); "El momento cientifico español 1775-1825," Madrid, C. Bermejo, 1934.
(5) Paneth, F., *Nature*, **159**, 8 (1947).
(6) Werner, A., "Neuere Anschauungen auf dem Gebiete der Anorganischen Chemie," 5th Auflage, p. 83, Braunschweig, Friedr. Vieweg & Sohn, 1923.

RECEIVED November 1951.

Chemical Nomenclature in Britain Today

R. S. CAHN and A. D. MITCHELL
The Chemical Society, London, England

> In spite of the common language, differences in the chemical nomenclature systems of Britain and America have developed. These differences lead to improper identification of compounds, confusion in indexing, and annoyances. These problems of conflict can be solved by close cooperation and consultation between British and American nomenclature policy makers.

Chemists will find it unsatisfactory that Mitchell (1) had to entitle his book "British Chemical Nomenclature." Because of the differences between British and American nomenclature, it is incorrect to maintain that there is at present only one chemical nomenclature in the English language. An examination of the nature of the differences and their causes will, it is hoped, help to remove these differences.

American developments of nomenclature reached their peak when British chemists were engrossed in World War II. One great American achievement was the "Ring Index," published in the year of Dunkirk. A second was the introduction to the index of *Chemical Abstracts*, 1945 and 1947. The lack of participation of the British in these publications was in part due to the war. No significant account of British practice was published until Mitchell's book (1948); until that time, nomenclature was not considered as a comprehensive, logically uniform whole. However, there are many points in which British nomenclature may be held to be superior to the American.

American and British Abstracts

In America the standard of nomenclature is set by the indexes to *Chemical Abstracts*. The individual abstract often uses the nomenclature of the original paper, which may be very different. Authors of papers in AMERICAN CHEMICAL SOCIETY journals are free, within limits, to use their own nomenclature, although they are encouraged to use the standard nomenclature. Editors of AMERICAN CHEMICAL SOCIETY publications, other than *Chemical Abstracts*, are not expected to supervise rigidly the smaller points of nomenclature used in papers in these journals.

Since Britain has a Society of Chemical Industry as well as a Chemical Society, *British Abstracts* was created in 1926 as a separate organization, which is independently administered by representatives of various societies. Since 1926, the (British) Chemical Society editors have no direct responsibility for nomenclature in *British Abstracts*. The Chemical Society sets the standard of nomenclature in Britain, but through its *Journal, Quarterly Reviews*, and *Annual Reports*, and not through abstracts or abstracts indexes.

Role of British Editors

The (British) Chemical Society is governed by an elected council, responsible for general policy. The council delegates to its Publication Committee responsibility for the policy of publications. The Publication Committee, and not the editor, makes the final decision for acceptance of papers and decides major matters of nomenclature. The editor is the executive officer of the Publication Committee. The editors (two of whom are full time and one part time) are regarded by the Publication Committee, and by the society as a whole, as expert advisers on nomenclature. It is the editors' duty and responsibility to ensure that correct nomenclature is used throughout the society's publications.

The editors make many nomenclature decisions on their own responsibility, but they will not alone introduce major changes. For advice on major questions, the Publication Committee appoints ad hoc subcommittees which reach limited objectives and are then disbanded. Standing subcommittees are not used. This difference in procedure is one of convenience. America has a large number of chemists, who can seldom meet and so work slowly. Because of shorter distances in the British Isles, committees can meet frequently and work rapidly.

The responsibility for British nomenclature is centered on the editor rather than on committees, and this affects the method of standardizing nomenclature. British nomenclature is designed primarily for written papers, American nomenclature primarily for indexes, and this produces many of the differences.

Nomenclature Differences

Nomenclature differences are of three types. One type causes real confusion, when one name indicates different substances in different countries or when different numbering systems are used. In the second type, the name is not ambiguous, but the position in indexes is different. For example, the British use glyoxaline and Americans use imidazole, and the British italicize prefixes such as *iso* while Americans do not. The third difference neither causes ambiguity nor affects the index, but is a nuisance—when colons are used between numerals instead of commas.

Descriptive Names. A good index name should also be a good name for cursive text. But a good index name is not always the most suitable for a particular context in a discussion. The question goes deeper than permission to use amino-alcohol or hydroxy-amine. A simple example is the acid I, which would be indexed as 2-ethylnonanoic acid. In a study of the influence of α-substituents in a series of acids it might be desired to name the acid as α-*n*-heptylbutyric acid or α-ethyl-α-*n*-heptylacetic acid. If the study concerned movement of the carboxyl group along the C_{10} chain, decane-3-carboxylic acid would be most appropriate. A chemist interested in fats might demand that the acid be named as a derivative of pelargonic acid. In appropriate circumstances, any of these names would be permitted in a British journal, though in most cases the Geneva name would be given at least once for the benefit of indexers.

C_7H_{15}
 CHCO$_2$H
C_2H_5

(CH$_3$)$_3$CCH$_2$
 CHCO$_2$H
(CH$_3$)$_3$CCH$_2$

I II III IV

An analogous case is the acid II, for which di*neo*pentylacetic acid was found more suitable in a descriptive paper than the systematic 2:2:6:6-tetramethylheptane-4-carboxylic acid or $\gamma\gamma$-dimethyl-α-*neo*pentylvaleric acid.

More complex cases are very frequent. The choice between 1-2'-thenoylphenanthrene, 2-1'-phenanthroylthiophen, and 1-phenanthryl 2-thienyl ketone for III, or between *p*-phenylbenzoic acid and diphenyl-4-carboxylic acid for IV illustrates the principle. The 1945 *Chemical Abstracts* index lists four methods of naming monoketones. Each method has its particular field of application. In cursive text it is not necessary to limit each field so strictly.

These examples illustrate the greater freedom available in cursive text. The need to choose one name for indexing remains, and the difference in names is not so much a conflict between text and index as a problem of how many cross references may be used in an index.

Substitutive and Additive Names. Real conflict does, however, arise in other ways. For instance, the bulk of classical nomenclature, certainly of International Union of Pure and Applied Chemistry rules, is based on substitution of groups into a parent compound by means of prefixes or one suffix—bromobenzene, 1-bromo-2-naphthylamine, phenylacetic acid. British chemists always write 3-indolylacetic acid for V. Indexers, however, wish to group indole derivatives together, so *Chemical Abstracts* uses the name indole-3-acetic acid. The orthodox "substitutive" reply is that grouping under an indole heading, if required, can be done as Indole, 3-carboxymethyl-. In a case as simple as V the difference is slight, for the substitutive name could be indexed beginning with Indolyl-, but the substitutive name for VI, α-bromo-β-3-indolyl-β-nitropropionic acid, is quite unsuitable for indexing. The name for VII [α-bromo-β-(5-bromo-3-indolyl)propionic acid] is no easier.

V   VI   VII

Consider the symmetrical compounds VIII and IX. Substitutive names might be methylenebis(aminoacetic acid) for VIII, and methylenebis(iminoacetic acid) for IX. Additive names might be methylenebis(iminoacetic acid) for VIII, and methylene(bis-nitriloacetic acid) for IX. The additive principle is not carried so far that the names begin with methanebis. The substitutive name for IX is the additive name for VIII, which is a most unpleasant situation.

VIII   IX

This difference reaches into inorganic chemistry. British organic chemists wish to call X methylphosphonic acid because the methyl group is substituted for hydroxyl in phosphoric acid XI rather than for hydrogen. By the same argument, XII should be fluorophosphonic acid, and XIII aminophosphonic acid. American practice is to call inorganic substances XII and XIII fluorophosphoric and aminophosphoric acid, respectively. (By an agreement which it is hoped shortly to publish, this discrepancy is being removed—July 9, 1952.) Many British inorganic chemists also see no objection to the phosphoric acid names, as they think that the substitutive principle is peculiar to organic chemistry.

X   XI   XII   XIII

Substitutive nomenclature is not convenient for polysubstituted compounds. XIV can be called 1:2-*cyclo*hexylenebisacetic acid, but there is no radical termination available for similar naming of the tris acid XV. This is similar to the situation whereby XVI is *o*-phenylenediamine, but XVII is benzene-1:2:3-triamine.

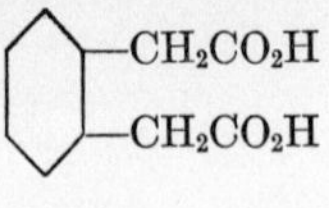

XIV 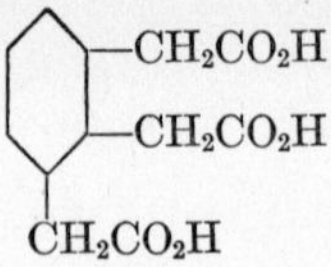 XV 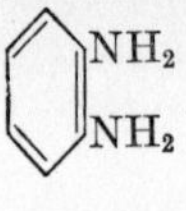 XVI 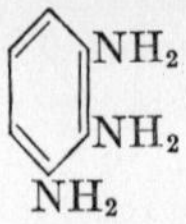 XVII

To follow substitutive nomenclature logically for complex cases, new polyradical terminations are required. The need for them, as well as the resultant troubles in indexing, are arguments against substitutive nomenclature. On the other hand, additive nomenclature is against the fundamental principles—*p*-hydroxybenzoic acid is not correctly described as water-benzoic acid or phenol-*p*-formic acid.

The convenience of indexers is not a primary consideration for nomenclature development. Here is a conflict requiring resolution. The obvious step is to define the limits within which the additive and the substitutive principle are severally preferable, unless all indexers adopt a ciphering system where such problems do not arise and the field is left clear for cursive text.

Cyclic Compounds. The Patterson-Capell "Ring Index" is the greatest single contribution to chemical nomenclature of recent years. Yet it has not been adopted *in toto* in Britain because some older methods are retained which are in general based on those of Richter. One principle, for instance, which is not favored is exemplified by the varied numbering of, say, the indene skeleton for which there

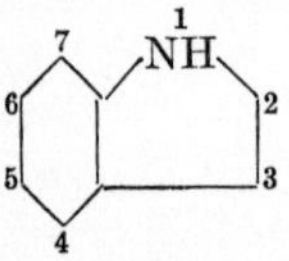

XVIII 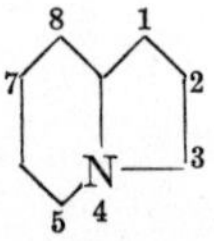 XIX 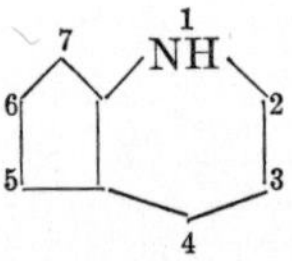 XX 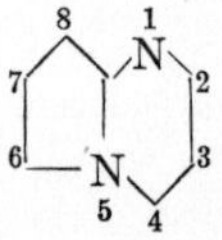 XXI

are four systems (XVIII–XXI). The rules for numbering atomic bridges but for omitting ring junctions leads to curious anomalies in complex cases. The partial numbering of the phenazine nucleus of XXII, and the irregular result at the side-chain junctions in XXIII are examples.

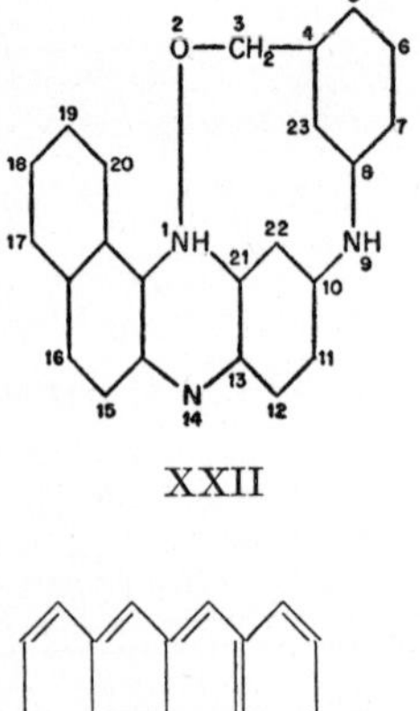

XXII 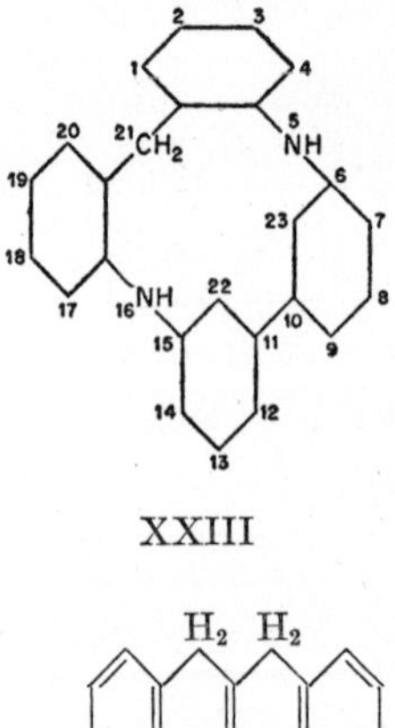 XXIII

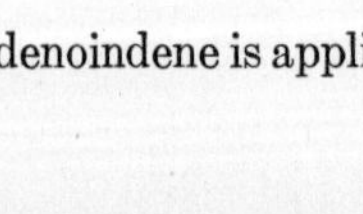

XXIV 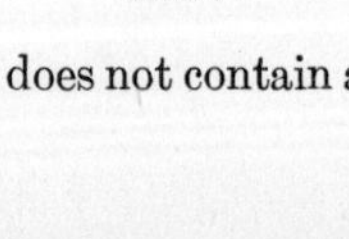 XXV

The name indenoindene is applied to XXIV which does not contain a CH_2 group characteristic of indene. The British use indenoindene for the substance XXV containing two complete indene nuclei, and name XXIV as a dehydro-derivative. There is real confusion when names such as indenoindene are applied by the Americans to the least hydrogenated form XXIV and by the British to the "typical" form XXV. There is conflict also in the cases where British numbering of rings differs from that used in the "Ring Index."

The "Ring Index" is the only work of reference in widespread use; it saves trouble for the author and provides a dictionary for the reader. However, final adoption of this system should await progress in the international deliberations on ciphering systems. It is desirable that any new method of naming and numbering rings should be compatible with the ciphering system eventually selected. Britain is reluctant to change nomenclature to the "Ring Index" system while this question is unsettled.

Minor Differences. Another point of difference between British and American custom is the use of di in diphenyl (Britain) and of bi in biphenyl (U.S.A.). The British do not use "double-molecule" names such as bibenzoic acid (dicarboxydiphenyl), because it is only the radical which is doubled. There is also the question of the state of reduction of cyclic ketones. The British define the state of reduction in the name of the compound, whereas American practice allows it to be taken for granted whenever possible.

Spelling differences—e.g., sulphur, ætio—are not important so long as they do not multiply. However, the alphabetical order in indexes is affected because the British italicize certain prefixes such as *iso* and *cyclo* which the Americans print in Roman type. Americans italicize sugar prefixes (*manno*saccharic acid); the British do not. Agreement should be reached on these points.

Cooperative Effort Needed

In relation to the whole field of nomenclature, these differences are not very important and can be readily adjusted. The similarities between the two systems exceed the differences. But it is not sensible that chemistry should be made harder than it already is by two English-language nomenclatures. The Chemical Society (*2*) has published the following statement:

The view was accepted by Council that international differences of nomenclature frequently form a real and serious barrier to the understanding of, and progress in, chemistry; that nomenclature, like grammar, has as its object the accurate and intelligible conveyance of information from one scientist to another; that no prestige or absolute value should attach to nomenclature apart from this object; and that, in consequence, conformity with internationally agreed nomenclature is most desirable even when the nomenclature so agreed may not seem to British chemists the best possible in particular cases.

It was decided that all feasible steps should be taken to foster international collaboration and to present British views for international consideration.

The British have adopted, from American practice, the alphabetical order of prefixes; have reconsidered part of the organophosphorus nomenclature in order to open the way to agreement in that field; and will make the many changes demanded by the 1949 IUPAC organic rules when these reach final form.

It is in keeping with the principles of IUPAC, if not indeed essential to them, that British-American agreement on broad questions should precede IUPAC discussions. There is in Britain today a desire to accord with international decisions and also a remembrance that British contributions to chemistry are still notable and that in nomenclature, too, our customs, though not immutable, should not be lightly cast aside. British experience is complementary to that of American authorities rather than competitive with it. Through the *Journal of the Chemical Society*, the editors are in close touch with authors, and are free to make changes at any time without regard to polyannual indexing periods. This is most important when a change as far-reaching as that of keto into oxo is involved, as it is in the 1949 IUPAC decisions. On the other hand, *Chemical Abstracts* has a wider range than one journal, and it has a nearly complete documented system.

The British and Americans each have an effective organization for considering and effecting changes. Though they operate differently, they can work in collaboration. The machinery of collaboration should be of the simplest type. The chief requirements are good will, the wish to agree, and the routing of committee consultations through the proper channels.

Editorial and intercommittee collaboration has already begun and positive results are beginning to accrue. The Board of Directors of the AMERICAN CHEMICAL SOCIETY has accepted a British proposal for a more formal recognition of this collaboration and has agreed with our Publication Committee that consultations shall precede the introduction of major changes in nomenclature by either society.

Literature Cited

(1) Mitchell, A. D., "British Chemical Nomenclature," London, Edward Arnold and Co., 1948.
(2) *Proc. Chem. Soc.*, (*London*) **1951**, 22.

RECEIVED August 30, 1951.

Chemical Nomenclature in the United States

E. J. CRANE

Chemical Abstracts, The Ohio State University, Columbus 10, Ohio

> Early and current efforts originating in the United States for standardization of chemical nomenclature on both a national and international level are reviewed. Committees sponsored by the National Research Council, which is the official organization for dealing with the IUPAC, and ACS committees correlate work to achieve consistency. These committees, with names and addresses of chairmen, are listed. British–U. S. cooperation on standardizing nomenclature, which resulted in joint adoption of ten rules in 1923, has been renewed. Nomenclature information is published in ACS periodicals and in eighteen pamphlets which are available from *Chemical Abstracts*.

Adam and Eve probably had a standard nomenclature. It was not chemical and very likely it was meager, but surely they agreed on the names "Cain" and "Abel" and probably on "apple" and "serpent." Yes, nomenclature begins in the beginning and this is true whether one is thinking of the beginning of human activities, the beginning of chemistry, or the beginning of pioneering work on a given chemical subject.

Little progress can be made in any activity without cooperation. Standardization of names or terms is necessary for efficient communication. Often, in new fields of endeavor, not enough is known to establish good nomenclature. Sometimes individuals grow fond of words coined under such circumstances and are reluctant to see them changed when fuller information dictates that change is necessary if the words are to fit into the framework of established nomenclature. For the good of a science, standardization must go forward and the group should rule over the individual in this respect.

This paper attempts to point to some of the early and recent efforts for standardization in chemical nomenclature originating in the United States and gives general information concerning some of the objectives, procedures, and accomplishments. Nomenclature workers sometimes deal with symbols, forms, abbreviations, and pronunciation, and their work sometimes includes trade names. This paper is limited largely to work on the naming of chemical compounds and elements.

Nomenclature workers should, above all others, use the word "nomenclature" correctly. It is sometimes used by groups to encompass all of the subjects just mentioned. This is not good usage. Incidentally this word is sometimes pronounced "nō'měn.klā'tûr" and sometimes "nō.měn'klà.tûr." Both are justified by Webster with "nō'měn.klā'tûr" given preference.

Achievements in Standardizing Nomenclature in the United States

A good deal of chemical nomenclature work has been done in the United States by organized groups. The AMERICAN CHEMICAL SOCIETY was organized in 1876; in 1886, just ten years later, an ACS Committee on Nomenclature and Notation stated: "Recognizing the importance of uniform practice in respect to nomenclature and notation among chemists speaking the same language, your committee has felt that none but the most weighty considerations would warrant any serious departure from the system already established by the English Chemical Society" (9). This committee then recommended that the English system, with minor modifications and additions, be adopted. This committee consisted of A. A. Breneman, G. E. Moore, A. R. Leeds, James H. Stebbins, Jr., and William Rupp. Two interesting observations by this group are:

1. It is more important that the vocabulary of a purely technical literature should be flexible and comprehensive than that it should be simple.
2. The present aspect of chemistry is such that a system of nomenclature that shall be at once comprehensive and permanent is practically unattainable.

What would these gentlemen think about the problem of finding a comprehensive and consistent nomenclature for chemistry today?

After four years of work a Committee on the Spelling and Pronunciation of Chemical Terms organized by the American Association for the Advancement of Science made an interesting report in 1892 (8). This committee consisted of T. H. Norton, Edward Hart, H. Carrington Bolton, and James Lewis Howe. The committee's report includes many good recommendations, some of which correspond to earlier British recommendations, but it also includes some rather radical proposals. Apparently these departures from usage have been regarded as made without sufficient reason for a change; they have never been accepted widely. One rule recommends the dropping of the final "e" in all chemical words ending in "-ine" to obtain such words as chlorin, amin, anilin, morphin, quinin, and cocain, and another rule recommends the dropping of the final "e" for chemical words ending in "-ide." In all cases it is recommended that a soft "i" be used in pronouncing the last syllable of these words. The Funk & Wagnalls Standard Dictionary has given preference to these recommended spellings, but they have never really "taken" in chemical circles.

This AAAS committee is, apparently, the only American group which has done work on the pronunciation of chemical words, except for the more extensive work on pronunciation done by the AMERICAN CHEMICAL SOCIETY's Committee on Nomenclature, Spelling, and Pronunciation (4). Individual British chemists were consulted when the ACS committee did its pronunciation work, but American usage was allowed to have a predominating influence.

Before the present ACS Committee on Nomenclature, Spelling, and Pronunciation was organized in 1911 there existed a Committee on Nomenclature and Spelling organized by the *Journal of the American Chemical Society*. From 1911 through 1946 the members of the board of editors of the *Journal of the American Chemical Society* always became members of the ACS committee. Beginning in 1911 and continuing to this day the editors of the journals published by the AMERICAN CHEMICAL SOCIETY have been members of its nomenclature committee, with the editor of *Chemical Abstracts* the chairman. Because of his pre-eminence in chemical nomenclature work it is appropriate that Austin M. Patterson should have been the individual who made the motion that the present ACS nomenclature committee be organized. He has long been a member.

Since this is in part an historical sketch, mention is made here of the interesting nomenclature suggestions of E. C. Franklin (10) for his liquid-ammonia work. We are accustomed to a so-called water world as far as chemical nomenclature and almost everything else are concerned. Franklin built up a corresponding nomenclature based on a liquid-ammonia world.

Besides the nomenclature work done by the general ACS nomenclature committee, sporadic early nomenclature work was done by committees of the ACS divisions. It was not until the need for more nomenclature activity, made evident in the chemical work done during World War II, was emphasized by Howard S. Nutting and Stewart S. Kurtz, Jr., that nomenclature work by committees of the divisions really began in earnest. Now there are numerous divisional nomenclature committees; a list is presented in Table I. The work on pectic substances by a committee of the ACS Division of Agricultural and

Food Chemistry reported in 1926 (*1, 7*) and work by a Committee on Labels of the Division of Chemical Education reported in 1932 (*2*) are examples of the earlier work of divisional committees.

NRC and ACS Nomenclature Committees. The U. S. National Research Council has sponsored nomenclature committees for specific purposes from time to time and continues to do so. Inasmuch as NRC is the official American organization for dealing with the International Union of Pure and Applied Chemistry, W. Albert Noyes, Jr., much interested in the work of both NRC and ACS, appointed Austin M. Patterson chairman of a general NRC chemical nomenclature committee in 1947 and asked that he organize subcommittees as needed. This step amounted to the reorganization of American committees under NRC for cooperation with IUPAC. Formerly there were independent American committees for organic chemistry, for inorganic chemistry, and for biological chemistry organized from time to time for this purpose.

The story of chemical nomenclature work in the United States would not be complete without a list of the present active nomenclature committees. These are given in Table I with the name and address of the chairman in each instance. These chairmen no doubt welcome any help offered.

The existence of so many nomenclature committees in the United States may seem puzzling to our colleagues overseas, particularly when they contemplate the need for consistency in the nomenclature of chemistry. These committees and their work are organized to facilitate cooperation. The nomenclature recommendations of the committees of divisions, which committees may have received reports from subcommittees, are always referred to the general ACS nomenclature committee for its consideration and action, after which they are referred to the ACS Council, which has the power to make rulings official for ACS. There is much overlapping of personnel in the ACS and NRC committees, with identical chairmen for certain branches of chemistry, which makes correlation of the work easy. All recognize a need for consistency so there is no real difficulty along this line.

It is clear that many are helping in the nomenclature work being done in the United States; many have helped. No one has contributed more than Austin M. Patterson, the dean of American chemical nomenclature workers. This active gentleman, experienced in dictionary work, has brought to the nomenclature task a high order of scholarship, a comprehensive knowledge of chemistry, a fair and judicious attitude, extraordinary ability to see the whole picture and thus keep decisions consistent, and a willingness to work for the common good. He has a full appreciation of the value of international nomenclature agreements, and he has participated in five IUPAC meetings. Dr. Patterson has been on IUPAC's Organic Commission since 1924.

Howard S. Nutting has been a dynamic factor in stimulating and organizing chemical nomenclature work in America and abroad in recent years. As chairman of the Committee on the Nomenclature of Organic Chemistry of the ACS Division of Organic Chemistry for the years 1946–51 he has brought nomenclature development and standardization forward rapidly.

Several members of the staff of *Chemical Abstracts* have done much to help in chemical nomenclature development, and they work continuously at tasks which further safeguard good usage. Of this group Leonard T. Capell is working part time on a straight nomenclature assignment. It would be good if we had a board of full-time nomenclature workers. Such a group could be kept busy to the distinct advantage of chemistry.

Nomenclature Rulings by Allied Organizations. Mention has been made of chemical nomenclature work within the AMERICAN CHEMICAL SOCIETY, of work by a committee of the American Association for the Advancement of Science, and of the activities of National Research Council committees. While most of the nomenclature work in chemistry is done by ACS and NRC groups, it would be a big mistake to assume that other American groups are not active in turning out nomenclature rulings of interest and value to the chemist.

The American Society for Testing Materials has a general Committee on Nomen

Table I. Active American Chemical Nomenclature Committees
(July 1952)

NATIONAL RESEARCH COUNCIL

General Committee on Chemical Nomenclature
    Howard J. Lucas, California Institute of Technology, Pasadena 4, Calif.
Subcommittee on Biochemical Nomenclature
    H. B. Vickery, chairman, Connecticut Agricultural Experiment Station, P.O. Box 1106, New Haven 4, Conn.
Subcommittee on Inorganic Nomenclature
    W. Conard Fernelius, chairman, Department of Chemistry, Pennsylvania State College, State College, Pa.
Subcommittee on Organic Nomenclature
    Howard S. Nutting, chairman, Dow Chemical Co., Midland, Mich.
Subcommittee on Steroid Nomenclature
    Byron Riegel, chairman, Department of Chemistry, Northwestern University, Evanston, Ill.
Subcommittee on the Nomenclature of Macromolecules (not under Lucas' NRC committee)
    Herman F. Mark, chairman, Brooklyn Polytechnic Institute, Brooklyn 2, N. Y.

AMERICAN CHEMICAL SOCIETY

General Committee on Nomenclature, Spelling, and Pronunciation
    E. J. Crane, chairman, *Chemical Abstracts*, The Ohio State University, Columbus 10, Ohio
Division of Analytical and Micro Chemistry
    S. E. Q. Ashley, chairman, Church Hill, Lenox, Mass.
Division of Organic Chemistry
    Mary Alexander, chairman, Universal Oil Products Co., 310 South Michigan Ave., Chicago 4, Ill.
    Subcommittee on Acids
        Charles D. Hurd, chairman, Northwestern University, Evanston, Ill.
    Subcommittee on Alcohols
        N. G. Fisher, chairman, Chemical Department, E. I. du Pont de Nemours & Co., Wilmington 98, Del.
    Subcommittee on Aldehydes and Ketones
        G. E. Goheen, chairman, 118 Parker Ave., Easton, Pa.
    Subcommittee on Definitions
        Mary A. Magill, *Chemical Abstracts*, The Ohio State University, Columbus 10, Ohio
    Subcommittee on Generic Terms
        Elmer Lawson, chairman, Patent Department, Sterling-Winthrop Research Institute, Rensselaer, N. Y.

---

clature and Definitions and many committees for specific subjects such as coal and coke and chemical analysis of metals. In particular, ASTM has defined carefully many technical terms used in industries. A considerable amount of standardization of terms of chemical interest has been accomplished by the American Standards Association. The ASA is now participating in work being done on nomenclature in the field of nuclear energy under the auspices of the National Research Council. Eventually this will be submitted to ASA for approval as "American Standard." Most ASA projects include definitions of terms. The American Institute of Chemical Engineers has been a useful contributor, particularly in relation to standard symbols.

Nomenclature rulings by groups which represent other sciences often interest the chemist. Those on applied spectroscopy by the Society of Applied Spectroscopy, those on colorimetry by the Optical Society of America, and those on symbols by the American Association of Physics Teachers are examples. The American Paper and Pulp Association, working in conjunction with the Institute of Paper Chemistry, has issued a "Dictionary of Paper," and the American Petroleum Institute will issue a "Glossary of Terms Used in Petroleum Refining." Further examples could be given.

Chemical Abstracts' Contribution. *Chemical Abstracts* and its staff have been in the midst of chemical nomenclature work in America since the beginning of this journal in 1907. Our task is to record the literature of chemistry, and a good recording job can be done only by careful attention to the use of good nomenclature. It is necessary, in particular, for a consistent systematic nomenclature to be used in the

Table I. Active American Chemical Nomenclature Committees (*Continued*)
(July 1952)

Division of Organic Chemistry (*Continued*)
    Subcommittee on Ethers
        Stanley P. Klesney, Dow Chemical Co., Midland, Mich.
    Subcommittee on Heterocyclics
        G. Dana Johnson, chairman, University of Indiana, Bloomington, Ind.
    Subcommittee on Hydrocarbons
        Louis Schmerling, chairman, Research and Development Laboratories, Universal Oil Products Co., Riverside, Ill.
    Subcommittee on Terpenes
        Mildred W. Grafflin, chairman, Research Department, Hercules Powder Co., Wilmington 99, Del.
    Advisory Committee on Nomenclature of Fluorine Compounds
        E. T. McBee, chairman, Department of Chemistry, Purdue University, Lafayette, Ind.
    Advisory Committee on Organophosphorus Nomenclature
        L. R. Drake, chairman, Physical Research Laboratory, Dow Chemical Co., Midland, Mich.
    Advisory Committee on Organosulfur Nomenclature
        F. G. Bordwell, Northwestern University, Evanston, Ill.
    Configurational Advisory Committee
        H. B. Vickery, chairman, Connecticut Agricultural Experiment Station, P.O. Box 1106, New Haven 4, Conn.
    Advisory Committee on the Nomenclature of Isotopically Labeled Organic Compounds
        Wallace R. Brode, chairman, National Bureau of Standards, Washington 25, D. C.
Division of Petroleum Chemistry
    Stewart S. Kurtz, Jr., chairman, Research and Development Department, Sun Oil Co., Marcus Hook, Pa.
Division of Physical and Inorganic Chemistry
    Committee on Inorganic Nomenclature
        W. Conard Fernelius, chairman, Department of Chemistry, Pennsylvania State College, State College, Pa.
    Committee on the Nomenclature of Physical Chemistry
        T. F. Young, chairman, Department of Chemistry, University of Chicago, Chicago 37, Ill.
Division of Rubber Chemistry
    Harry L. Fisher, chairman, 200 North Wayne St., Arlington, Va.
Division of Sugar Chemistry and Technology
    M. L. Wolfrom, chairman, The Ohio State University, Columbus 10, Ohio

---

building of indexes, and *Chemical Abstracts* has emphasized the importance of publishing thorough, scientifically prepared indexes. The indexes to *Chemical Abstracts* are sometimes used in the United States as a word-by-word guide to good chemical nomenclature, and index introductions, particularly that for 1945 mentioned below, are a source of general nomenclature information.

While the *Chemical Abstracts* indexes are used widely as sources of nomenclature information, it is pointed out that names systematized for indexing are not necessarily the best names for all purposes. *Chemical Abstracts* does not require that index names be used in abstracts, but often they are so used to advantage.

The "Directions for Abstractors and Section Editors of *Chemical Abstracts*" contain nomenclature rules as well as other information of general interest in chemical writing, and these are used by other than *Chemical Abstracts* workers. When *Chemical Abstracts* was organized it borrowed information from British directions, particularly nomenclature information.

ACS Committee on Nomenclature, Spelling, and Pronunciation. Since the editor of *Chemical Abstracts* always has been the chairman of the ACS Committee on Nomenclature, Spelling, and Pronunciation, it is sometimes difficult to distinguish between nomenclature work done for the committee and for *Chemical Abstracts*.

Developmental work in chemical nomenclature, limited at times, is far from being the only work done by the ACS nomenclature committee. At the present time our committee conceives that it can best be useful by (1) channeling problems which arise so that they receive consideration by the proper group of specialists, (2) helping to keep the rulings of the special committees consistent with the general nomenclature picture, (3) eventually bringing their recommendations to the ACS Council for official approval, (4) with NRC cooperation, seeking international acceptance of adopted national standards, (5) widely distributing nomenclature information, with some care to get it into strategic places to improve usage, (6) answering the many queries received concerning the application of existing nomenclature rules, (7) attempting to discourage bad practices, (8) endeavoring to keep decisions by nonchemical groups in accord with chemical nomenclature rulings when these involve chemical compounds, as when entomologists name insecticides, and (9) attempting to help authors and editors when nomenclature proposals are offered for publication. Our cooperation with nonchemical groups has been extensive. It has included work for such groups or undertakings as the American Medical Association, the U. S. Pharmacopeia, Merck's Index, the Chemical Rubber Handbook, the Interdepartmental Committee on Pest Control, the publishers of dictionaries, and editors of other journals.

Space will not permit recapitulation in this paper of nomenclature accomplishments by the ACS Committee on Nomenclature, Spelling, and Pronunciation. Reports have appeared annually, for many years in the *Proceedings of the American Chemical Society*, and, in recent years, in *Chemical and Engineering News*. Of interest are ten rules (see page 62) agreed upon as a matter of British-American cooperation and a pronunciation survey. The work on the nomenclature of the hydrogen isotopes and their compounds (*5*), that on the written form (*6*), and that on silicon compounds (*3*) are also noteworthy. A committee headed by Wallace R. Brode is now working on the nomenclature of compounds containing isotopes other than those of hydrogen.

Publicizing Nomenclature Rulings. Publication of nomenclature information has been a problem. Should proposals be published promptly or should an effort be made through committees to determine general acceptability? The former may lead to helpful discussion, but it also may lead to confusion in the literature from too early acceptance. Awaiting a committee report complicates the crediting of suggestions. The modern tendency is to favor going slow on publication. *Chemical and Engineering News'* recent inauguration of a regular nomenclature column, conducted by Austin M. Patterson, is an interesting and promising development in nomenclature publication.

Standard nomenclature rules are of limited value if many chemists are not familiar with them. As chairman of the ACS committee the author endeavors to place special emphasis on the wide distribution of nomenclature information. In addition to the writing of numerous letters thousands of copies of the following nomenclature pamphlets have been distributed:

1. Definitive Report of the Commission on the Reform of the Nomenclature of Organic Chemistry. Translation with comment and index by Austin M. Patterson. 10 cents.
2. Rules for Naming Inorganic Compounds. Report of the Committee of the International Union of Chemistry for the Reform of Inorganic Chemical Nomenclature, 1940. The committee has provided an American version of the rules. 10 cents.
3. The Pronunciation of Chemical Words. A committee report. 5 cents.
4. Nomenclature of the Hydrogen Isotopes and Their Compounds. A committee report. No charge.
5. Directions for Abstractors and Section Editors of *Chemical Abstracts*. Much concentrated information on nomenclature, symbols, forms, and abbreviations is assembled in this 46-page booklet in form convenient for use. 25 cents.
6. The Standardization of Chemical Nomenclature. This reprint of an article by the committee chairman contains a list of references to sources of information on chemical literature. No charge.
7. The Naming and Indexing of Chemical Compounds by *Chemical Abstracts*. Introduction to the 1945 Subject Index. A comprehensive, 109-page discussion of chemical nomenclature as applied to inorganic as well as organic compounds for systematic indexing, with a classified bibliography, an index, and the following appendixes

(lists): (I) Miscellaneous chemical prefixes, (II) Inorganic groups and radicals, (III) Anions, (IV) Organic groups and radicals, (V) Organic suffixes, and (VI) 1945 ring index. 75 cents.

8. The Nomenclature of Silicon Compounds. A committee report. No charge.

9. The Nomenclature of the Carotenoid Pigments. Report of the Committee on Biochemical Nomenclature of the National Research Council, accepted by the Committee on Nomenclature, Spelling, and Pronunciation of the AMERICAN CHEMICAL SOCIETY. No charge.

10. Nomenclature of Natural Amino Acids and Related Substances. A committee report. 10 cents.

11. Carbohydrate Nomenclature. A committee report. 10 cents.

12. The Naming of Cis and Trans Isomers of Hydrocarbons Containing Olefin Double Bonds. A committee report. No charge.

13. The Designation of "Extra" Hydrogen in Naming Cyclic Compounds. A committee report. No charge.

14. Avogram. A committee report. No charge.

15. The Naming of Geometric Isomers of Polyalkyl Monocycloalkanes. A committee report. No charge.

16. Commission de nomenclature de chimie inorganique. This relates to the names and symbols of the elements of atomic numbers 4, 13, 41, 43, 61, 71, 72, 74, 85, 87, 91, 93, 94, 95, and 96. In English. 10 cents.

17. Commission de nomenclature de chimie biologique. This includes rules for the nomenclature of natural amino acids and related substances (in English) (these will no doubt, when completed, supersede 10 above) and Nomenclature de vitamines (in French). 10 cents.

18. Commission de nomenclature de chimie organique. This includes rules on the nomenclature of organosilicon compounds (some modifications of 8 above, which they supersede, are included), changes and additions to the Definitive Report, extended examples of radical names, and an extensive list of radical names. All in English. 50 cents.

Copies can be obtained from the office of *Chemical Abstracts*, The Ohio State University, Columbus 10, Ohio.

Over 7500 copies of the Definitive Report (1, above) have been sold and the Naming and Indexing book (7 above) has gone over 6300 copies in sales. The pamphlets listed are a mixture of International Union reports (sometimes with added notes and examples), ACS committee reports, and *Chemical Abstracts* publications.

Highlights in American chemical nomenclature activity have been (1) the adoption in 1915 of systematic naming, with inverted organic entries, for the indexing of chemical compounds by *Chemical Abstracts*, (2) the publication in 1940 of "The Ring Index" by Austin M. Patterson and Leonard T. Capell (this activity was sponsored jointly by the National Research Council and the AMERICAN CHEMICAL SOCIETY), and (3) the publication in 1945 by *Chemical Abstracts* of "The Naming and Indexing of Chemical Compounds by *Chemical Abstracts*." The last two are very useful publications.

The *Chemical Abstracts* treatise on naming and indexing and Mitchell's "British Chemical Nomenclature" (*11*) are the most comprehensive publications on chemical nomenclature in existence.

Achievements in Standardizing Nomenclature on an International Basis

While much nomenclature work has been done in the United States without participation by chemists of other countries, there has never been a time when there was not readiness by most American nomenclature workers to enter into international negotiations on nomenclature problems. Americans are not a particularly patient people, and sometimes the delays naturally necessary for international agreement have been allowed to count as an obstacle to cooperation. At other times war conditions have interfered. Nevertheless, Americans have participated actively in the nomenclature work of the International Union of Pure and Applied Chemistry; international rulings have been adopted promptly and widely for use in this country; and many of our rulings have been submitted for international consideration with our subsequent acceptance of recommended alterations.

Perhaps it should be mentioned that rulings by the International Union of Pure and Applied Chemistry have not been accepted on one or two occasions. For example, some IUPAC biochemical nomenclature recommendations made in 1923 were protested both in this country and in England, and these were withdrawn. This sort of action has been rare.

British-American Cooperation. To a degree, of course, there are language barriers in international nomenclature standardization. This is not true in the case of our British colleagues, and it is only natural that there should have been more British-American cooperation than cooperation by the United States with any other individual country. Specific examples will be discussed later. An early example of cooperation with another nation was an agreement entered into in 1902 by several American chemists to help the Society of German Engineers compile a trilingual (German, English, French) "Technolexicon."

Mention has been made of consideration given to British nomenclature rulings. The chairman of the ACS nomenclature committee sought in 1919 to make British-American cooperation definite, and he was supported in this by a resolution passed by the ACS Council. It reads: "That the President of the AMERICAN CHEMICAL SOCIETY invite on behalf of the Council of the Society the governing bodies of the Chemical Society (London) and the Society of Chemical Industry to appoint a committee, or committees, on Nomenclature, Spelling, and Pronunciation to cooperate with the corresponding committee of the AMERICAN CHEMICAL SOCIETY in order to secure as large a measure of agreement in these fields as is practical." This led to the joint British-American adoption in 1923 of ten nomenclature rules covering some of the more commonly disputed points. Some of these relate to endings which have a classification significance. These rules, which are useful to this day, follow:

1. In naming a compound so as to indicate that oxygen is replaced by sulfur the prefix *thio* and not *sulfo* should be used (sulfo denotes the group SO_3H); thus, HCNS, *thio*cyanic acid; H_3AsS_4, *thio*arsenic acid; $Na_2S_2O_3$, sodium *thio*sulfate; $CS(NH_2)_2$, *thio*urea. The only use of *thio* as a name for sulfur replacing hydrogen is in cases in which the sulfur serves as a link in compounds not suitably named as mercapto derivatives; thus, $H_2NC_6H_4SC_6H_4NH_2$, thiobisaniline. *Hyposulfurous acid*, not hydrosulfurous acid, should be used to designate $H_2S_2O_4$.
2. The word *hydroxide* should be used for a compound with OH and *hydrate* for a compound with H_2O. Thus, barium hydroxide, $Ba(OH)_2$; chlorine hydrate, $Cl_2.10H_2O$.
3. Salts of chloroplatinic acid are *chloroplatinates* (not platinichlorides). Similarly salts of chloroauric acid are to be called *chloroaurates*.
4. Hydroxyl derivatives of hydrocarbons are to be given names ending in -*ol*, as glycer*ol*, resorcin*ol*, pinac*ol* (not pinacone), mannit*ol* (not mannite), pyrocatech*ol* (not pyrocatechin).
5. The names of the groups NH_2, NHR, NR_2, NH, or NR should end in -*ido* only when they are substitutents in an acid group, otherwise in -*ino*; thus, MeC(:NH)OEt, ethyl im*ido*acetate; $H_2NCH_2CH_2CO_2H$, β-am*ino*propionic acid (not am*ido*propionic acid); $PhNHCH_2CH_2CO_2H$, β-anil*ino*propionic acid; $CH_3C(:NH)CO_2H$, α-im*ino*propionic acid.
6. Hydroxy-, not oxy-, should be used in designating the hydroxyl group; as *hydroxy*acetic acid, $HOCH_2CO_2H$, not *oxy*acetic acid. *Keto-* is to be preferred to *oxy-* or *oxo-* to designate oxygen in the group —CO—.[a]
7. The term *ether* is to be used in the usual modern acceptation only and not as an equivalent of *ester*.
8. Salts of organic bases with hydrochloric acid should be called *hydrochlorides* (not hydrochlorates nor chlorhydrates). Similarly hydrobromide and hydroiodide should be used.
9. German names ending in -*it* should be translated -*ite* rather than -*it*, as per-

---

[a] The name oxo- for —CO— was approved by IUPAC in 1930 and is now used by *Chemical Abstracts* and by many others in the United States. However, the name keto- has never been completely outlawed; in 1949 it was recommended that keto- be restricted in use to the general sense, as in ketohexose.

mut*ite*. If it seems desirable to retain the original form of a trade name it should be placed in quotation marks, as "Permutit." Alcohols such as dulcitol (German Dulcit) are exceptions.

10. German names of acids should generally be translated by substituting *-ic acid* for "*-säure.*" Some well-established names are exceptions, as Zuckersäure (saccharic acid), Milchsäure (lactic acid), or Valeriansäure (valeric acid), etc. For a few well-established names it is correct to translate "*-insäure*" *-ic acid* instead of *-inic acid*. E.g., Acridinsäure is acridic acid. Names ending in "carbonsäure" are to be translated *-carboxylic acid* (not *carbonic acid*).

In connection with this cooperative work, the author corresponded with British colleagues, in particular with A. J. Greenaway in the early days. He edited abstracts in England from 1885 to 1921 and the *Journal of the Chemical Society* during the period 1921–23. He seems to have played the leading part in the preparation of British rules for abstracting, which included nomenclature rules. Relations among British and American editors have remained most pleasant. There has been a recent renewal of effort for cooperation in nomenclature; the central British figure in this renewal is R. S. Cahn, editor of the *Journal of the Chemical Society*. Corresponding with him on nomenclature work has been a satisfying experience. Previously wartime conditions had made such activities difficult, yet these conditions emphasized the need for more nomenclature work. In recognition of this situation a resolution was passed by the ACS Board of Directors in June of 1951 as follows: "It was moved, seconded, and carried that in so far as possible, the AMERICAN CHEMICAL SOCIETY and the Chemical Society (London) cooperate in the standardization of chemical nomenclature, each participant to be free to act independently whenever local conditions demand immediate action, and that the chairman of the Committee on Nomenclature, Spelling, and Pronunciation be authorized to work out details with The Chemical Society. The foregoing action formalizes a practice which now is being followed and restates an action by the Council taken on September 2, 1919." The purpose is agreement among English-speaking peoples on nomenclature with the desirability of a more general international agreement kept in mind constantly. More than half of the scientific literature of the world is now published in the English language.

Perhaps it is not out of place here, since British-American relations have been discussed, to mention that Austin M. Patterson and the present speaker were invited to write the chapter on "Nomenclature, Chemical," of the last edition of "Thorpe's Dictionary of Applied Chemistry" (*12*). This was done.

Nomenclature development is slow work; the workers are usually busy men with much else to do. Sometimes there is need for early decisions. These facts have, to a degree, been stumbling blocks in the attaining of international agreement. Nomenclature work within a nation is much prolonged when put on an international basis. Usually, however, the effort is worth while. It is not practicable and, perhaps, not desirable for everything in the nomenclature field to be done on an international basis, as some more or less specific problems are largely local in interest.

Significance of Standardization

Interest in chemical nomenclature in the United States is now high. In earlier days only a limited number of individuals took interest. In some new fields of activity even now nomenclature interest lags at first, but picks up rather early. It was difficult in the early days to persuade industry to care about the nomenclature used. Now industrial chemists are in the forefront among those who are active in nomenclature standardization. They recognize that the existence and use of good, standard chemical nomenclature rules are truly important—for effective communication, dependable record keeping, and the saving of much effort and time. Exactness in the reporting of scientific information is important just as is exactness in the work done in the laboratory.

Literature Cited

(1) AMERICAN CHEMICAL SOCIETY, Committee for Revision of the Nomenclature of Pectic Substances, *Chem. Eng. News*, **22**, 105–6 (1944).

(2) AMERICAN CHEMICAL SOCIETY, Committee on Labels, *J. Chem. Education*, **7**, 2937–42 (1930).
(3) AMERICAN CHEMICAL SOCIETY, Committee on Nomenclature, Spelling, and Pronunciation, *Chem., Eng. News*, **24**, 1233–4 (1946).
(4) AMERICAN CHEMICAL SOCIETY, Committee on Nomenclature, Spelling, and Pronunciation, *Ind. Eng. Chem., News Ed.*, **12**, 202–5 (1934).
(5) *Ibid.*, **13**, 200–1 (1935).
(6) AMERICAN CHEMICAL SOCIETY, Committee on Nomenclature, Spelling, and Pronunciation, *J. Chem. Education*, **8**, 1336–8 (1931).
(7) AMERICAN CHEMICAL SOCIETY, *J. Am. Chem. Soc.*, **49**, Proc. 37–9 (1927).
(8) AMERICAN CHEMICAL SOCIETY, *Proc. Am. Chem. Soc.*, **1892**, 63–70.
(9) Chemical Society, *J. Chem. Soc.*, **41**, 247–52 (1882).
(10) Franklin, E. C., "The Nitrogen System of Compounds," AM. CHEM. Soc. Monograph No. 68, New York, Reinhold Publishing Corp., 1935.
(11) Mitchell, A. D., "British Chemical Nomenclature," London, Edward Arnold and Co., 1948.
(12) "Thorpe's Dictionary of Applied Chemistry," London, Longmans, Green, and Co., Ltd.

RECEIVED July 1951.

Basic Features of Nomenclature in Organic Chemistry

FRIEDRICH RICHTER
Beilstein-Institut, Frankfurt a. M., Hoechst, Germany

> Chemical nomenclature, though it is a prerequisite of the science, is not perfectly consistent and logical because it has developed as a language for the communication of chemical knowledge. This language contains speech elements rooted in trivial or common names, and established by long usage and association. Characteristic aspects of organic nomenclature and their relation to history are discussed in this paper. Systematic and nomenclatural needs are difficult, but not impossible, to reconcile. Rules and definitions can be imposed on a nomenclature system only within the limits allowed by a growing language.

Nomenclature is written in bold characters over the gateway through which the domain of modern chemistry is entered. In the introduction to his famous "Traité élémentaire de Chimie" of 1789 Lavoisier (*38*) wrote the following passage:

When I undertook writing this work I had nothing in mind but expanding my paper on the necessity of improving chemical nomenclature. During this work, I felt more than ever the evidence of the principles laid down by the Abbé de Condillac (*17*) in his Logic by stating that we think only with the help of words, and languages are veritable analytical methods. And as a matter of fact, while I believed myself concerned with nomenclature only there grew between my hands this elementary treatise quite unexpectedly and without my being able to do anything about it.

These sentences will serve as an introduction to the study of nomenclature. They direct attention to two fundamental aspects, the close connection of nomenclature with the state of knowledge, and the character of nomenclature as a language. Scientific language, embracing terminology as well as nomenclature, is thought of as an artificial language, created according to the need of naming the subject matters studied and of describing them in a way that makes their salient features appear related against a background of basic, systematically connected ideas. When the word "language" is pronounced, perspectives are opened up into the field of linguistics also. The language aspect, turning up wherever thought finds expression, is inseparable from the historic growth process of nomenclature. This coexistence of logic and pragmatic aspects is the reason why nomenclature, though a prerequisite of science, is lacking in perfect logical consistency and borders on what is figuratively called an "art."

In all of the outstanding synopses of nomenclatural rules, only part of the whole story has been given, and another part, equally essential for understanding what nomenclature actually is, has remained undiscussed. It has not been realized how much advantage can be gained from the acknowledgment of certain basic ideas which give a less random direction to practical efforts toward perfection and, above all, give the public a better approach to what has often appeared as a puzzling and disturbing field.

All nomenclature in chemistry started from "common" or "trivial" names, a term the origin of which goes back to Linnaeus (*42*). "Systematic" or "rational" nomenclature, as Tiemann (*57*) has expressed it, aims at "spoken formulas." From a phenomenological point of view, it is highly interesting that, as far as the roots of the different "speech elements" in nomenclature are concerned, systematic nomenclature still leans heavily on designations of trivial origin. The logical character and the capacity of nomenclature for conveying meaning reside on what may be called its "grammar"—i.e., on the arrangement of the single constituents, called speech elements, on rules of "inflection" by endings, etc. Much can be achieved by such a simple means, especially if the unique problems offered by organic chemistry are considered, such as the description of structure in terms of geometric patterns, the complication of which is still increasing daily with the progress of science. It is due primarily to the simple laws which govern this architecture that a nomenclature of the familiar form has given reasonably satisfactory results.

Classification of Names

In a broad sense, four general types of names may be distinguished in organic nomenclature:

1. Functional names proper (type names)
2. Substitution names
3. Additive names
4. Replacement names (thio- and a- names)

Organic language rests, to a large measure, on functional and substitution names. Different trends in the development of these types may be traced back in history.

History of Functional and Substitution Names. The basis of systematic nomenclature was laid in the early thirties of the nineteenth century. Then, Lavoisier's hypothesis of radicals ("compound elements" as he defined them) as constituent parts of organic acids received experimental verification by Liebig and Woehler's (*40*) paper on the "radical of benzoic acid," where the persistence of a radical called "benzoyl" in diverse chemical transformations was proved. A considerable body of other evidence was soon to be interpreted in a similar way. Liebig and Woehler introduced the ending -yl for radicals, deriving it from Greek ὕλη = matter. Later, ethyl was derived from ether (*39*), methyl from the entity methylene coined by Dumas and Péligot (*20*) in their paper on wood spirit after μέθυ = mead and ὕλη = wood, thus giving the etymology of the ending -yl a curious ambiguity (*10*).

The creation of systematic signification by attaching endings to trivial names, here in its infancy, was prophetically expressed by Dumas (*18*) writing with respect to camphene:

I proposed therefore for this hydrocarbon the ending *-ene* in order to avoid confusion with the alkaloids. This is necessary since, at least for a long time, the entire art of organic nomenclature will consist in modifications of endings.

The first use of radical names was not in substitution names but in "functional names proper." Based on the theories of types which stressed structural analogies to the binary inorganic compounds, these names referred organic compounds to types or classes—e.g., chlorides, alcohols, ethers, ketones, sulfides, etc. As a rule, the class characteristic was the name of an individual compound promoted to serve as prototype, and to this end specified by the name of the respective radical. Dumas and Berzelius generalized the term alcohol when methyl alcohol was so named in order to infer the analogy with alcohol. Acetone had stood for a class name a long time when homologs of it were prepared. Though the term "ketone" had been introduced by Gmelin (*25*) as early as 1848, the class term "acetones" can occasionally be found in the nineties.

All of the functional names proper have this in common: the terminal does not signify an individual compound into which something is substituted, but a type according to which the compound is built. They may thus also be called type names. The introduction of the important term "function" is credited to Gerhardt (*24*). He writes:

One may derive chemical compounds from a certain number of typical formulas; one thus groups together all of the acids, then all of the ethers, etc. One may, in chemistry as in plant physiology, consider all trees indiscriminately, the relation of the leaves to each other, then of the flowers, etc.; such is in chemistry the classification according to types or functions.

The experiences which found expression in Dumas' "substitution theory" led to a second type of names. The sensational observation of hydrogen replaced by other elements without fundamental change of type was immediately visualized by Dumas (19) as having a bearing on nomenclature. Lavoisier's binary nomenclature was now no longer sufficient. Dumas writes:

It is necessary that each type have a name, that this name be conserved in all of the numerous modifications it can undergo. On this principle I have already formed the names acetic and chloroacetic acid, ether and chloroether—names with an aim at recalling the persistence of types in spite of the intervention of chlorine in these compounds.

The modern practice of symbolizing substitution by placing the respective designations before the otherwise unaltered name of the parent compound grew from this approach. The field for substitution names widened beyond limits when, with the inauguration of structural chemistry and the aromatic theory, a detailed picture of the chemical formula became available and hydrocarbon radicals now competed with inorganic substituents in producing an unforeseen multiplicity of names.

Functional names proper built from hydrocarbon radicals and type-denoting terminals are now restricted to simple compounds belonging to alcohols, ethers, ketones, sulfides, and amines. The majority of names, however, are now built by extension of the substitution principle, inasmuch as not only prefixes but also suffixes are visualized as introduced into the parent compound by way of substitution of hydrogen. This has come about by a process of blending typical of the ways of nomenclature. The systematic names of carboxylic acids and sulfonic acids were doubtless functional names proper originally. The "carbonic acids" as they are still called in German after the model set by Kolbe (36) were regarded as analogs of carbonic acid with one hydroxyl replaced by hydrocarbon radicals, and, for a long time, methyl sulfonic acid was the preferred name for the now orthodox methanesulfonic acid (11, 33, 41, 61).

Principles of Present-Day Usage. Of expressions coming under the heading terminology, the "speech elements" or "word elements" form the smallest building blocks, and the "parent compound" is the basis or stem of the name which is modified by the word elements in the way of affixes. Technically, the modification of the parent compound is considered as a substitution of hydrogen atoms by side chains, as characterized by the ending -yl in the simplest case, on the one hand and what is often called "functional" and "nonfunctional substituents" on the other hand. (In Beilstein terminology, the corresponding terms are "Funktionen" and "Substituenten.") Some difficulty is experienced with the definition of the term "substituent," partly owing to confusion of the "definiens" and the "definiendum." In the teaching of chemistry, the term "substituents" connotes the actual groups relevant with respect to classical substitution theory—i.e., the halogens and the nitroso, nitro, azido, and sulfo groups. The "definiendum substituent" in nomenclature, however, implies the side chains plus functional plus nonfunctional substituents, which is unfortunate.

The situation is no better with regard to the "function" introduced into the official rules of the Geneva convention (16, 57) in 1892 without definition and which, theoretically, might imply any group distinct by its chemical reactivity including double and triple bonds (26). Judging from predominating usage, the term refers to those groups for which prefixes as well as suffixes are available. Practically, this would agree with the Beilstein definition (52), the only one based on a structural criterion, according to which the presence of hydrogen linked to an inorganic atom and therefore available for derivative formation defines the function. As a subgroup, Béhal (7) has introduced the term "derivative function" for ethers, sulfides, etc. The general subject of terminology in nomenclature is in need of reform.

Nomenclature of Parent Compounds

The main body of rules of nomenclature usage refers to the standardization of designations for radicals, functions, and substituents, provisions for their arrangement, and definition of the range where they may be applied. According to whether the affixes thus defined are combined with trivial or systematic names of parent compounds, semisystematic or systematic names result. Though not unexpected for the specialist, it is surprising to find that truly systematical names for parent compounds are restricted to a few types. The saturated aliphatic hydrocarbons are designated by Greek and Latin numerals with the ending -ane preconized by Hofmann (29) in 1865. They are interesting because they usher in a tradition of tacit implication—neither carbon nor hydrogen is explicitly mentioned. The modification of the alkanes by the endings -ene and -ine (or -yne) in order to signify unsaturation is a most convenient device of Hofmann. It has been made official by the Geneva rules. Analogical naming of hydroaromatic compounds by addition of the prefix cyclo to the systematical names of the alkanes has also been introduced at Geneva.

Fused Ring Compounds. For aromatic rings, trivial names like benzene and toluene gained stature by being given the rank of official names. The variety of polycyclic compounds is, however, too great for trivial names as the only basis of naming. Hantzsch (27) has introduced a useful method of naming bicyclic ortho-condensed systems of aromatic saturation by a device called "fusion" ("Anellierung" in German). The process is most easily visualized if naphthalene is thought of as the result of superimposure of two benzene rings with two "congruent" 1,2 positions and ensuing fusion of the two. This procedure is expressed by the speech equivalent benzo-benzene. The process is symbolized by attachment of the ending -o to the name of the fused hydrocarbon, though modification of these prefixes for reasons of euphony is frequent (9).

The application of the fusion principle allows minor variations with respect to the stage of hydrogenation. American practice (48) of referring always to the lowest possible stage of hydrogenation has much to recommend it. For systems predominantly saturated, Baeyer (6) has invented the system of "bicyclo-names" which is founded on the idea of "bridge heads." It is especially useful when more than two consecutive ring atoms are common to both rings. These names are extensions of the Geneva cyclo names, carrying the prefix "bicyclo" and a "characteristic" enclosed in brackets. Bicyclo[3.3.1]-nonane symbolizes a saturated system of 9 carbon atoms built in such a way that bridges of 3, 3, and 1 carbon atom each connect 2 carbon atoms selected as bridge heads. To a certain measure, the practice is even suitable for extension to tricyclic systems, though additional rules for the choice of the primary bridge heads are required (14, 43).

The special geometry of aromatic systems allows fusion at more than two neighboring atoms only in the form of "multiple ortho-condensation," no two rings having more than two atoms in common. Of the "reticular systems" thus resulting, peri-condensed systems are the simplest and most important ones. Names for them can be formed by the fusion principle if it is kept in mind that an additional "operation in H" is required to make up for the missing hydrogen atom at the quaternary carbon atom common to the three rings. Comparison of the ortho-condensed 1,2-benzanthracene ($C_{18}H_{12}$) and the peri-condensed 1,9-benzanthracene ($C_{17}H_{12}$) shows that if peri-fusion is permitted, the prefix "benzo" is not a fixed entity unless the special kind of operation is further determined by the locants. On the whole, the principle of "grafting," as the fusion may also be visualized, has often presented a convenient solution for naming complex systems.

Heterocyclics. In the field of heterocyclics (9), still bigger problems turn up which are usually disposed of by a lavish assortment of trivial names. How fruitful a simple chemical idea may prove for the creation of semisystematic nomenclature is shown by Knorr's (35) suggestion to name pyrazole after pyrrole, -azole to signify replacement of a CH group by nitrogen. Hantzsch (28) thereby was inspired to characterize heterocycles of 5 ring members by the ending -ole and designation of the heteroatoms by prefixes like oxo, thio, imido, etc. Names like oxdiazole, triazole, thiodiazole, etc., have since then been in common use. Widman (58) has set the fashion for 6-membered rings by exactly analogous use of the ending -ine, as in triazine. Polycyclic heterocycles can often be dealt with by the fusion principle. A wide range of application is open to the important "*a* nomenclature" simultaneously developed by Stelzner (53) and Patterson (46) from a happy blending of earlier suggestions by Bouveault (12), Ingold (31), and Sudborough (55). Here, heteroatoms qualified by the ending -a as in oxa, thia, aza, etc., are prefixed to the respective carbon analoga of equal ring size. Pyridine and piperidine would be called azabenzene and azacyclohexane according to this principle.

Because of the dislike of endings not in harmony with chemical behavior and a decisive preference for shortness, an alternative system of naming has been suggested for the monocyclic heterorings other than 5 and 6 rings by Patterson (46). By modification of the respective Greek numerals, the syllables "ir" and "et" for 3 and 4 rings, and "ep," "oc," "on," and "ec" for 7 through 10 rings have been proposed in the same manner for use as -ole in 5 rings. This gives azetidine for azacyclobutane, azocine for azacyclooctatetraene. Simplification of the provisions for designation of the different hydrogenation stages would be a precondition to widespread use.

Numbering

Uniqueness of names formed according to the rules is attained only if a system of numbering is agreed upon. This task is a particularly arduous one, the importance of which has been grasped only slowly, and, therefore, the number of undisputed principles in this field is small. Among them is the rule that in hydrocarbon chains, in order to locate modifications, the chain is numbered by arabic numerals beginning at one end, in such a way that the "locant" of a modification becomes as low as possible. Whenever an order of seniority of the modifications has to be established, the principles diverge. The Geneva rules and with them the "Beilstein Handbook" give highest rank to the carbon side chains and among them to the smallest one. The Liége rules, on the contrary, give the highest rank to the "principal" function, without specifying the order of seniority.

In cyclic compounds, where the ring often plays the role of the main chain, the ring atom getting number 1 is found according to the same rules with appropriate modifications. For side chains, the most usual system of numbering consists in numbering them de novo and giving 1 to the point of attachment at the chain or the ring, respectively.

The principle of giving the lowest locants to the simplest or what is considered the most important modifications has often been applied to trivial names. Whenever the starting point is unambiguous as in toluene, phenol, or benzoic acid, where the ring atom connected with the modification is numbered 1, the success is satisfactory. However, the confusion possible with trivial names containing several substituents is familiar from the cresols and toluidines. There is no reason why an agreement between these numbering principles should not be reached.

For symmetrically substituted systems, aliphatic as well as noncondensed aromatic ones, the system of priming has usually given satisfactory results. It is in the field of condensed polycyclics that a unified solution is especially difficult to attain. For the simpler polycyclics like naphthalene, anthracene, quinoline, etc., conventional all-round numberings of long standing exist, which have formed the model for other systems. In the early developments, numbering of the fused positions seemed uninteresting since hydrogenation and synthesis of hydroaromatics played a minor role. Since extension not guided by general principles only led to confusion, it appeared as a gratifying aspect of the fusion names that they numbered the components of the fused systems separately, according to the familiar numberings of the smaller units (9, 54). The locants of the parent compound are then made recognizable by not being primed. Though the system is reliable, the numbering is often unwieldy and bound to the particular type of fusion name.

Patterson (45) has devised workable rules for an all-round numbering of condensed systems. The system evolves from the sound concept that a convention for uniform drawing of structural formulas has to precede numbering. He has further lightened the burden of necessary numbers by denoting the fused sides of the parent compounds by lower-case letters, putting a equal to 1,2 and then traveling around the perimeter. The starting point of the clockwise numbering is defined as the "first free angle in the right upper quadrant." Carbon atoms in the fusion positions are numbered by affixing a to the preceding locant. This system is used in *Chemical Abstracts* and applied by Beilstein whenever no conflicting policy is in the way. Its usefulness is greatest with reticular systems. In all instances where the system is applicable and trivial names are still favorite, the unambiguous numbering provided by it is especially valuable since, with trivial names, the "pons asinorum"—i.e., possibility of reconstructing the numbering system by help of locants of characteristic groups—is usually lacking.

Interest in a completely systematic naming and numbering of polycyclic condensed compounds has been revived by the invention of new ciphering systems. By the joint effort of Taylor (*56*), Patterson (*47*), and Dyson (*21*), ring size of the constituent unsaturated rings is expressed by tetralene, pentalene, hexalene, etc., the number of rings by the distributive Latin numerals bini, terni, quaterni, etc.; thus anthracene is ternihexalene. In contradistinction to the older systems, all positions without exception are numbered starting from a fusion position and entering the next ring via the lowest locant of the preceding one. The locants of "overstep" are combined into a characteristic which defines the exact kind of fusion. Disadvantages of this system are the complete departure from all tradition, irregularities of sequence, and certain difficulties for the inexperienced in locating the starting point which, though unique, is often found only by trial and error. This is a field where discussion has only just begun, and all potentialities will have to be weighed carefully. This system would place naming and numbering of the polycyclics on an equal footing with the simpler compounds.

Radicals, Functions, Substituents

The names for hydrocarbon radicals are governed by simple and fairly consistent rules with which the public is generally familiar. On the contrary, little of systematic significance is to be found when the names of functions and substituents are considered. For the most part, they are taken from the names of the elements or from trivial designations sanctioned by tradition. It is patent that the intervention, in the functions, of inorganic elements with varying valence and unsuitability for application of the substitution principle make them mostly the field of opportunism.

Nitro was coined by Mitscherlich (*44*) in his work on benzene. Nitroso stems from Church and Perkin's (*15*) putative "nitrosophenylin" and Laurent's expression "substitutions nitrosées." Nitrosophenylin proved later to be aminoazobenzene and did not contain a nitroso group at all. Azobenzene got its name by Zinin's (*62*) French translation of Mitscherlich's name for it "Stickstoffbenzid." Azoxy was another designation by Zinin. Wurtz (*59*) introduced amino in 1849. Hofmann coined hydrazo after azo (*30*). Hydrazine and hydrazone were derived by Fischer (*23*) from hydrazo. Baeyer introduced carboxy (*4*) in 1865, keto (*5*) in 1886. A misunderstanding prompted Kekulé to reject keto and recommend oxo in its place (*1, 2, 37*). The prefix hydroxy, derived from the term hydroxide, for designation of alcohols and phenols is of very old standing. The corresponding -ol is connected with alcohol and further entrenched by Wurtz's (*60*) glycol. The first official mention of -ol as an ending of alcohols and phenols occurs in an article by Armstrong (*3*) in 1882 where he polemized against the German habit of calling benzene "benzol" and stated that -ol for alcohols and phenols was recommended by the London Chemical Society. This seems to be the origin of the later Geneva rule. The disregard of the identical ending -ol in heterocyclics like pyrrol(e) was glossed over by an orthographic artifice neither very convincing nor generally applicable. It is difficult to avoid mistakes in the (often pointless) attempt to rationalize trivial names without direct structural connotation. Jacobson (*8*) stressed this a long time ago.

As a general ending for ketones the suffix -one is very old. It is the ending of acetone that has stood sponsor to most of them, beginning with Péligot's (*51*) "benzon" which Chancel (*13*) later baptized benzophenone. The origin of the analogous name "acetophenone" is not quite clear. It seems to have sprung from Baeyer's laboratory in 1870 (*22*). The orthodox definition of -one implies replacement of H_2 by $=O$ in a nonterminal position. A second practice, the thrust of which has been underrated by the Geneva convention, started in the middle of the eighties from v. Pechmann's (*49*) pyridone and Knorr's (*34*) pyrazolone. By these names it is tacitly understood that preceding hydrogenation is required for rendering introduction of the oxo group possible. The coexistence of these conflicting usages (which occasionally have been extended to aromatic hydrocarbons with still additional specifications) illumines the tendency for short designations "at any cost," and the extreme difficulties of a consistent integration in the field of nomenclature.

Nomenclature as a Language

Nomenclature is a language aimed at the communication of chemical knowledge in an optimal way as to conciseness, associative suggestivity, and systematic significance, in accord with the subject matter under investigation. The desire of the individual to make himself understood in the most satisfactory way has its objective counterpart in the social aspects of nomenclature, which therefore exhibits some features of common language. A basic feature of their relation to common language is that individuals are born into it and receive it from a social group. This is true of chemical language, too, even though it is for the most part an artificial creation. In order to become nomenclature, words or rules must be "accepted." Nomenclature is not unified and integrated from the start. Its historical growth may be compared to a process of crystallization, starting simultaneously from different centers and ending in lines of discontinuity where areas originally separated come into mutual contact. Thus, collection and acknowledgment of good usage capable of generalization are a continuous social process.

Nomenclature, as a language, has for its objective the easy flow of thought in speech as well as writing, and will try to reach this goal by all means within its compass. This is illustrated by the coining of new trivial designations in spite of the availability of correct systematic names. It is also obvious in certain synthetic features reminiscent of the concept of "holism" (*Ganzheit*) in linguistics, in that not the "speech elements" but only the "syntax"—i.e., the way the portions are joined together—decides the meaning. The different implications of names like hexanone and pyrazolone are such a case. Likewise, in thiodiacetic acid and thiobenzoic acid, thio means different things, understood only from the context. These examples are mentioned to illustrate the actual life of nomenclature. The conciseness of short designations for complicated structures and the method of tacit inference, called the synthetic feature, are life elements of nomenclature which will maintain themselves. As a scientific tool, nomenclature should be made as consistent and systematic as possible, but there are limits in the pursuit of this goal where the specialist finds himself in the role of a gardener carefully pruning a plant.

Nomenclature and Systematics

What relation exists between nomenclature and systematics? The question is by no means as tautologic as it appears. Systematics is defined as "integrated classification"—i.e., institution of agreement between different principles of classification so that every chemical species is allocated in it a unique or nearly unique place. In the architecture of names, the systematically relevant portions are usually concentrated in the end or at best in the middle of the names. This is the reverse of what is required by systematics based primarily on functions. If the Geneva rules are regarded as an attempt to reconcile nomenclatural and systematic needs, the putting in relief of the hydrocarbon portion as the systematically most relevant portion must appear as a natural consequence. By beginning with the stem and suffixing the most important functions in a fixed order, the Geneva names have a distinctly systematic flavor which has made them attractive for use in "Beilstein's Handbook."

In order to take advantage of this systematic effect in indexing, measures have to be taken which are opposed to the natural tendencies of nomenclature. The device known as "inversion" has often been considered in connection with the Geneva nomenclature and hexane, methyl- as an alternative to methylhexane may be cited as an example. The decision to make this principle a general policy in the subject index to *Chemical Abstracts* may be characterized as a retrieval of nomenclature for indexing purposes. Generally speaking, nomenclature is far too roughly classificatory and overlapping to be helpful in producing a systematic arrangement. An attempt to make nomenclature as systematic as possible by breaking down the structural formula to the smallest portions that may be depicted by nomenclature leads back to the "official names," the drawbacks of which were discussed fifty years ago by Jacobson and Stelzner (*32*) in a paper still worth studying.

Certain ciphering systems, by giving linear representations ("notations") of structures in terms of (mostly familiar) partial structures, are, to a certain measure, only nomenclature further abbreviated. The suggestion has been made of retranslating ciphers into words as a step toward a much standardized nomenclature. The real problem, however, is not the construction of official names, but the coexistence of official and conversational names. The overstandardized approach inherent in ciphering and official nomenclature is too narrow to satisfy all needs of chemical thinking, and conversational language will always take the first place in the mind of the productive chemist. There is, however, still another aspect to the problem of retranslation of ciphers. The number of usable word elements and rules is small. Attempts to find new solutions by new permutations of the same word elements should be avoided as confusing and an imposition on the memory of the chemist.

Any future modification of nomenclature should be in the direction of fortification rather than destruction of such logic as there is in present nomenclature. Translated ciphers can hardly fulfill this requirement with the help of the traditional word treasure. Too much is required from nomenclature as a medium of systematization. Systematics have remained a concern for editors of abstract journals and handbooks only and are remote from the minds of the public. If systematics were given their proper part instead of heaping all of the burden on nomenclature, progress in the future might be easier.

Nomenclature and Structure Formulas

A close relation between name and formula is revealed by Tiemann's definition of Geneva names as "spoken formulas." The symbolic nature of both is evident in the way they persist more or less unchanged in spite of deep changes due to evolution of the ideas connected with them. The meaning of a word may be dim in the origin, the number of associations it carries may increase with growing experience, and inadequate associations may be subdued or dropped, but the word itself, because it is not too specific, will survive and the continuous change of meaning will pass nearly unnoticed. Chemical names like formulas depict chemical structure only in terms of sequence of atoms connected by bonds in a way satisfying conventional valency requirements. Questions relating to character and strength of bonds and their gradation are customarily neglected. Designations like chlorides and oxides, originally borrowed from ionic inorganic compounds, continue in use. The name acetylene is still used though Berthelot derived it from acetyl which he believed to be C_2H_3. The problem of "aromatic character" is veiled under the merciful cloak of Mitscherlich's name benzene as well as under the conventional formula depicting only sigma bonds.

With the rise, in recent times, of electronic chemistry and wave mechanics, more detailed information has been incorporated into our formulas in terms of octet symbols, fractional charges, etc. The functional and abstract character of all wave-mechanical representations and the modern interpretation of many chemical formulas as "limiting structures" under the aspect of complementarity do not make it probable that nomenclature will receive much stimulation and development from this angle.

The Future

The increasing agreement on principles of nomenclature, owing to international cooperation, is gratifying, and it is to be hoped that, in the future, it will be extended to the greatest possible number of details. There is still a superabundance of variants as to numbering, proper place of the locants in the names, limits of rules set by the national languages, problems of appropriate translation, etc. A careful revision and statement of the logical principles underlying chemical nomenclature would be of great educational value. The formal tools needed for the enunciation and exposition of the wealth of practical and theoretical knowledge of modern science are of prime importance. May the study of nomenclature be guided by the words of the philosopher C. S. Peirce (50):

> Woof and warp of all thought and research are symbols, and the life of thought and science is the life inherent in symbols; so that it is wrong to say that a good language is important to good thought, merely, for it is the essence of it.

Literature Cited

(1) Anschuetz, R., "August Kekulé," Vol. I, p. 587, Berlin, Verlag Chemie, G.m.b.H., 1929.
(2) Anschuetz, R., and Parlato, E., *Ber.*, **25**, 1977 (1892).
(3) Armstrong, H. E., *Ibid.*, **15**, 200, footnote (1882); *J. Chem. Soc.*, **41**, 247 (1882).
(4) Baeyer, Adolf von, *Ann.*, **135**, 307 (1865).
(5) Baeyer, Adolf von, *Ber.*, **19**, 160 (1886).
(6) *Ibid.*, **33**, 3771 (1900).
(7) Béhal, A., *Bull. soc. chim.*, (4) **11**, 271 (1912).
(8) "Beilstein's Handbuch der organischen Chemie," 4th ed., Vol. V, p. 6, Berlin, J. Springer, 1922.
(9) *Ibid.*, 4th ed., Vol. XVII, p. 7, Berlin, J. Springer, 1933; 2nd suppl., Vol. XVII, p. 3, Berlin-Goettingen-Heidelberg Springer-Verlag, 1952.
(10) Berzelius, J. J., letter to Liebig, Dec. 19, 1834, "Aus Methy und Hylae Methylène zu machen, ist ganz sprachwidrig," in "Berzelius und Liebig. Ihre Briefe von 1831–1845," 2nd ed., p. 96, by J. Carrière, Muenchen, J. F. Lehmann, 1898.
(11) Blomstrand, C. W., "Chemie der Jetztzeit," p. 157, Heidelberg, 1869.
(12) Bouveault, L., Assoc. franc. pour l'avancement des sciences, Congrès de St. Etienne, 1897.
(13) Chancel, G., *Ann.*, **72**, 280 (1849).
(14) *Chemical Abstracts*, **39**, 5885, introduction to Subject Index (1945). Naming and Indexing of Chemical Compounds.
(15) Church, A. H, and Perkin, W. H., *Quart. J. Chem. Soc.*, **9**, 1 (1857).
(16) Combes, A., in "Dictionnaire de Chimie," Vol. C, 2nd suppl., p. 1060, by C. A. Wurtz, Paris, Hachette et Cie., 1894.
(17) Condillac, E. B. de, "La logique, ou les premiers développements de l'art de penser," Paris, 1781; Le Roy, G., "Oeuvres Philosophiques de Condillac," Vol. II, p. 371, Paris, Presses Universitaires de France, 1948.
(18) Dumas, Jean, *Ann.*, **9**, 64 (1834).
(19) Dumas, Jean, *Compt. rend.*, **10**, 168 (1840).
(20) Dumas, Jean, and Péligot, E. M., *Ann. chim.*, (2) **58**, 5 (1835); *Ann.*, **15**, 1 (1835).
(21) Dyson, G. M., "A New Notation and Enumeration System for Organic Compounds," 2nd ed., p. 29, London, Longmans, Green and Co., 1949.
(22) Emmerling, A., and Engler, Carl, *Ber.*, **3**, 886 (1870).
(23) Fischer, Emil, *Ibid.*, **8**, 589 (1875); **21**, 984 (1888).
(24) Gerhardt, Charles, "Traité de Chimie Organique," Vol. IV, p. 611, Paris, F. Didot, 1853; "Précis de Chimie Organique," Vol. I, p. 5, Paris, 1844.
(25) Gmelin, L., "Handbuch der Chemie," Vol. IV, 4th ed., pp. 120, 181, Heidelberg, K. Winter, 1848.
(26) Grignard, V., "Traité de Chimie Organique," Vol. I, p. 1074, Paris, Masson & Cie., 1935.
(27) Hantzsch, Arthur, and Pfeiffer, G., *Ber.*, **19**, 1302 (1886).
(28) Hantzsch, Arthur, and Weber, H. J., *Ibid.*, **20**, 3119 (1887).
(29) Hofmann, A. W., *J. prakt. Chem.*, (1) **97**, 272 (1866).
(30) Hofmann, A. W., *Proc. Roy. Soc. (London)*, **12**, 576 (1863); *Jahresber.*, 424 (1863).
(31) Ingold, C. K., *J. Chem. Soc.*, **125**, 88 (1924).
(32) Jacobson, Paul, and Stelzner, R., *Ber.*, **31**, 3368 (1898).
(33) Kekulé, F. A., "Lehrbuch der Organischen Chemie," Vol. II, p. 499, Erlangen, F. Enke, 1866.
(34) Knorr, Ludwig, *Ann.*, **238**, 145 (1887).
(35) Knorr, Ludwig, *Ber.*, **18**, 311 (1885).
(36) Kolbe, A. W. H., *Ann.*, **101**, 264 (1856).
(37) Kolbe, A. W. H., *J. prakt. Chem.*, (2) **2**, 390 (1870).
(38) Lavoisier, A. L., "Traité élémentaire de Chimie," nouvelle ed., p. V, Paris, 1789.
(39) Liebig, Justus von, *Ann.*, **9**, 18 (1834).
(40) Liebig, Justus von, and Woehler, Friedrich, *Ibid.*, **3**, 249 (1832).
(41) Limpricht, H., and Uslar, L. v., *Ibid.*, **102**, 248 (1857).
(42) Linné, Carl von, "Philosophia botanica, in qua explicantur fundamenta botanica," p. 202, Stockholm, 1751.
(43) Mitchell, A. D., "British Chemical Nomenclature," p. 107, London, Ed. Arnold and Co., 1948.
(44) Mitscherlich, E., *Poggendorf's Ann. Physik.*, **31**, 625 (1834).
(45) Patterson, A. M., *J. Am. Chem. Soc.*, **47**, 543 (1925); *Rec. trav. chim.*, **45**, 1 (1926).
(46) Patterson, A. M., *J. Am. Chem. Soc.*, **50**, 3076 (1928).
(47) Patterson, A. M., private communication, 1947. Possibilities for a Combined System of Notation and Nomenclature for Organic Compounds.

(48) Patterson, A. M., and Capell, L. T., "Ring Index," p. 22, New York, Reinhold Publishing Corp., 1940.
(49) Pechmann, H. v., *Ber.*, **18**, 318 (1885).
(50) Peirce, C. S., "Collected Papers," ed. by Hartshorne, Charles, and Weiss, Paul, Vol. II, p. 129, Harvard University Press, 1931-35.
(51) Péligot, E. M., *Ann.*, **12**, 41 (1834).
(52) Prager, B., Stern, D., and Ilberg, K., "System der organischen Verbindungen," p. 9, Berlin, J. Springer, 1929.
(53) Stelzner, R., "Literatur-Register der organischen Chemie," Vol. V, p. IX, Berlin-Leipzig, Verlag Chemie, G.m.b.H., 1926.
(54) Stelzner, R., and Kuh, E., "Literatur-Register der organischen Chemie," Vol. III, p. 21, Berlin-Leipzig, Verlag Chemie, G.m.b.H., 1921.
(55) Sudborough, J. J., *J. Indian Inst. Sci.*, **7**, 181 (1925).
(56) Taylor, F. L., private communication, 1947. Proposed System of Enumerative Nomenclature for Organic Ring Systems.
(57) Tiemann, Ferd., *Ber.*, **26**, 1621 (1893).
(58) Widman, O., *J. prakt. Chem.*, (2) **38**, 185, 189 (1888).
(59) Wurtz, C. A., *Compt. rend.*, **29**, 169 (1849).
(60) *Ibid.*, **43**, 200 (1856).
(61) Wurtz, C. A., "Dictionnaire de Chimie," Vol. III, p. 120, Paris, Hachette et Cie., 1874.
(62) Zinin, N. N., *J. prakt. Chem.*, (1) **36**, 93 (1845).

RECEIVED November 1951.

Organic Chemical Nomenclature, Past, Present, and Future

P. E. VERKADE
Laboratory of Organic Chemistry, Technical University, Delft, Netherlands

> Trivial names are used profusely in organic chemistry, and they are essential. The Definitive Report of the Commission for the Reform of the Nomenclature of Organic Chemistry, the basis of present systematic nomenclature, does not forbid the use of trivial names. Rule 1 states that "as few changes as possible will be made in terminology universally adopted." The introduction of new trivial names, however, requires careful consideration. Systematic nomenclature began in 1892 with the establishment of the Geneva Rules by a committee appointed by the International Chemical Congress of 1889. The international work practically came to a standstill until the International Union of Pure and Applied Chemistry was formed in 1919. Since then impressive results have been obtained in establishing a systematic nomenclature in spite of numerous problems.

In the course of the century and a half of the official existence of their science—Berzelius in the beginning of the 19th century introduced the name organic chemistry, which has since gradually acquired its present meaning—organic chemists have for only a few decades enjoyed the unlimited right of bestowing on substances isolated or synthesized by them names of their own choice (trivial names).

Trivial Names

A very great number of trivial names are used, or have been used in organic chemistry, and this number is still increasing, day by day. In their quest for names, organic chemists have always shown great ingenuity. By inventing impressive names with Latin or, more often, Greek roots, they proved that their classical education was not wasted. Sometimes they even allowed a glimpse of their thinking or of their feelings: a striking example of such a case is barbituric acid, provided it is true that this is named in honor of a certain Barbara.

It is really remarkable that the origin of many generally accepted and constantly used trivial names is practically unknown. For example, it would be interesting to find out the percentage of organic chemists who are familiar with the origin of the name "phene," the use of which as a synonym for benzene was sanctioned in rule 13 of the Definitive Report of the Commission for the Reform of the Nomenclature of Organic Chemistry of the International Union of Chemistry (IUC), adopted in 1930 (11). This name can still be recognized in terms like phenyl, phenylene, and phenol, but is never used as such. The study of the origin of trivial names, and especially of the many obsolete names, is fascinating and amusing. Moreover, it is without doubt useful, because historical research of this nature can lead to clarification in other fields.

As a whole, the Definitive Report mentioned above, which is the basis of our present nomenclature, deals with the method of forming systematic names. This has sometimes led to the idea that the use of trivial names should be limited in the extreme; for instance, that names like acetylene and acetic acid should be replaced by ethyne and ethanoic acid (methane carboxylic acid). The Netherlands Patent Office used to go rather far in this direction, and instruction in organic chemistry in Dutch secondary schools also tends to accept such a point of view. This attitude is not correct. It is absolutely contrary to the spirit of the Definitive Report; indeed, rule 1 says that "as few changes as possible will be made in terminology universally adopted," and surely it may be assumed that this applies to current trivial names.

In the present state of organic chemical nomenclature trivial names are absolutely indispensable. Examples in support of this assertion are plentiful; what should we do without names like glucose, saccharose, streptomycin, penicillin, quinine, and strychnine? It is out of the question that in this respect the situation will become different in the near future, and it seems justifiable to question whether trivial names will ever be abandoned.

A question that will always require careful consideration, however, is whether introduction of a new trivial name is justified. Besides, replacing certain trivial names by more or less systematic ones can result in greatly improved surveyability of various highly specialized parts of organic chemistry and thus facilitate their further development. An example is the excellent report on the nomenclature of steroids published a short time ago in various languages (1) as a result of a conference of specialists in this field; with the aid of a limited number of conventions, it appeared possible to replace the cumbersome trivial names of many compounds of well-known structure by systematic names of very reasonable length and form from which the structure can clearly and easily be deduced.

Systematic Names

Systematic names are those that give a more or less complete idea of the structure of the compounds in question—starting from a certain number of trivial terms, of course. In the course of the second and third quarters of the 19th century, the accumulating insight into the structure of organic compounds rendered systematic naming possible in a rapidly increasing degree. The enormous increase in number of these compounds, the result of the admirable development of synthetic methods, necessitated the use of such names whenever possible. During the International Chemical Congress held in 1889 in Paris, a committee was set up to put a stop to the confusion prevailing already at that time in the realm of systematic nomenclature, by making concrete proposals. The result of the work of this committee, which consisted of representatives from various European countries, was the establishment of the Geneva Rules in 1892 (13). What may be called "directed nomenclature" was thus introduced into organic chemistry.

After the acceptance of the rules just mentioned, international work on nomenclature practically came to a standstill until the International Union of Pure and Applied Chemistry (IUPAC) was formed in 1919. This organization appointed a Commission for the Reform of the Nomenclature of Organic Chemistry; the work of this Commission, carried out under Holleman's leadership from 1924 to 1930, led to the adoption of the already mentioned Definitive Report (11) containing the Liége Rules, a considerably improved version of the Geneva Rules. Since then this international commission, later named the Commission on the Nomenclature of Organic Chemistry, has put forward a number of improvements and additions to the Liége Rules; these will be referred to later in another connection.

The results of 60 years of international activity in the field of organic chemical nomenclature are perhaps not apt to make much impression. However, it should be borne in mind that work of this kind is very difficult and requires great diplomacy, because very often there is the problem of reconciling fundamentally divergent points of view, especially those of the editors of the paramount national periodicals and handbooks. Moreover, during the greater part of the period in question there was, as a whole, little or no interest in problems of nomenclature; this assertion is not rendered invalid by the fact that Istrati published a book of not less than 1223 pages in Romanian on the subject in 1913, or that the Italian pharmacist Siboni for some years edited a private periodical called *La Nomenclatura Chimica*, which was mainly devoted to advertising a useless system of nomenclature invented by himself (*9*).

Lately, a remarkable change with regard to the attitude toward nomenclature problems may be perceived in some countries, particularly in the United Kingdom and in the United States. Especially in the latter country the nomenclature-mindedness has increased a great deal. Ever more groups of organic and other chemists turn their interests toward such problems, and it is clear that this stimulates to a marked degree the work of the International Commission. The recent drastic revision of the rules for radical names (*7*) and the recent rules on the nomenclature of organosilicon compounds (*6*) are mainly the work of American organizations. It is to be hoped that such an interest in problems of nomenclature may soon develop in many other countries belonging to various linguistic realms.

It is a striking fact that many of those who find the courage to tackle the study of nomenclature problems soon become very much interested in the subject and fond of the work. This is a denial of the current opinion that work of this kind is dull. Here, also, there is much truth in the saying, "unknown, unloved." Still, even if the work on nomenclature problems were far from entertaining, it would have to be done, as it is of such great importance to the development of chemistry as a whole and of organic chemistry in particular.

Need for True International Cooperation

Rule 1 of the Geneva Rules expressed the wish that authors would make a habit of mentioning in their papers the official systematic names of the compounds in brackets after the names of their choice. It would not have been unreasonable to go somewhat further and express the wish, or even the expectation, that the new nomenclature should be applied as much as possible. The Liége Rules do not contain any text of this nature, and this is certainly not because it was deemed superfluous at the time. Unfortunately, even now it cannot be said that it is general practice to endeavor to adhere to the rules. Only too often serious deviations from the rules are encountered, when these are not in the least necessary, and now and then more or less official new agreements, contrary to the Liége Rules, have been made without the consent or even knowledge of the International Commission. A striking example of this deplorable fact follows.

The discussions leading to the establishment of the Liége Rules were attended at various times by Professor Grignard, who was not a member of the International Commission. This chemist expressed opinions and submitted proposals which were eventually not incorporated in the Definitive Report. Yet Grignard carried through these rejected ideas and many others, often absolutely contrary to the Liége Rules, in a place where he was almighty. For the "Traité de Chimie Organique" (*4*) which, under his editorship, started to appear in 1935, he created what, without much exaggeration, may be called a nomenclature of his own. It is quite obvious that this could only harm the cause of international systematic nomenclature in France.

The strong support of the national organizations adhering to the Union and of the editors of periodicals, abstract journals, and handbooks is absolutely essential for an adequate rationalization of organic chemical nomenclature.

Flexibility of Systematic Nomenclature

The Definitive Report allows the user great liberty in certain respects, even if its rules are carefully followed. This fact may be illustrated by the following examples.

According to the Geneva Rules, the carbon atom of the carboxyl group of aliphatic acids is regarded as part of the fundamental chain, so that, for instance, the systematic name of acetic acid was to be ethanoic acid. Long discussions preceded the adoption of the rules in question. A comparatively large minority of those present at the Geneva conference preferred to take a different course, namely, to consider the carboxyl group as a substituent, in which case the systematic name of acetic acid would be methane carboxylic acid. During the meetings of the Commission for the Reform of the Nomenclature of Organic Chemistry, another endeavor was made to win a victory for the latter principle, which, at least at the time, was applied rather generally and which was obviously the only solution when dealing with carboxyl groups bound to ring systems (benzene carboxylic acid). All arduous efforts in this direction failed, especially as a result of the strong opposition of Professor Grignard. The duels in words between him and Professor Holleman, then the president of the International Commission, will not be forgotten by those who attended the meetings in question. It was a jolly moment when Holleman pointed out that it was Grignard himself who had found the most illustrative reaction in support of the consideration of the carboxyl group as a substituent, namely, the action of carbon dioxide on an organomagnesium halide. The confusion became even worse when the old Professor Henry Armstrong, who was not a member of the Commission and had never concerned himself with its business, came in and immediately suggested a third possibility for the nomenclature of aliphatic acids, according to which, for instance, acetic acid would have to be called methanoic acid. The result was that the Geneva nomenclature was maintained, on the understanding that in cases where the use of that nomenclature would not be convenient, the substituent principle could be applied. This, of course, also applies to the numerous types of derivatives of the carboxylic acids in question. As opinions may differ considerably on the question of whether a name or a principle is convenient or not, the rule in question actually leads to an enormous number of compounds having two official systematic names.

In rule 51 of the Liége Rules it is stipulated that for compounds possessing different kinds of functions only one kind (the principal function) will be expressed by the ending of the name, whereas the other kinds of functions will be designated by appropriate prefixes. The choice of the principal function, however, is left entirely open, although it appears from unpublished preliminary drafts of the Definitive Report that at first it was proposed to set up a definite order of functions. For example, the compound $CHO.CH_2.CH_2OH$ can, therefore, equally well be called 3-hydroxypropanal or 3-oxo-1-propanol. It is obvious that in complicated cases the lack of regulation of the choice of the principal function must lead to a considerable variety of possible systematic names.

In rule 6 of the Liége Rules, which deals with saturated aliphatic hydrocarbons with side chains, definite instructions are given for naming these compounds. However, if a simpler name would result, another prescribed way of naming should be used. The simplicity of a name is obviously very much a question of personal opinion. Consequently, in this instance also, more than one method of naming is officially tolerated.

Finally, it may be remarked that according to rule 7 of the Liége Rules the side chains of hydrocarbons can be designated in order of complexity or in alphabetical order. Here again there is a choice between two possibilities of naming, which, moreover, is complicated by the fact that the alphabetical order is not the same in all languages.

Revision and Extension of the Definitive Report

The International Commission has made revision and extension of the Definitive Report a part of its program and has decided to accomplish this difficult task in sections. A start has been made recently with the treatment of the material under the headings "generalities" and "hydrocarbons." It is obvious that in this respect very good use will be made of the wealth of material brought together in the papers delivered at the "Joint Symposium on the Nomenclature of Hydrocarbons," held in Atlantic City in 1949 (8).

The question now arises as to whether liberties such as those mentioned are to be allowed in the forthcoming version of the Definitive Report. On the one hand, a certain flexibility of nomenclature may sometimes be of considerable usefulness; for example, the free choice of the principal function in compounds with various kinds of functions makes it possible to choose a systematic name which emphasizes the type of research carried out with the compounds in question. On the other hand, such liberties form a serious difficulty in the compiling and using of indexes; they constitute one of the reasons why different indexes rest on a different basis. It is obvious that this fact renders efficient literature searching rather difficult. The disadvantages are undoubtedly greater than the advantages in the instance, and, therefore, it is necessary to aim at such a wording of the future report that the rules always lead to unique names.

All over the world the liberty of the individual is constantly decreasing; directed economy binds people more and more, and in ever new ways. Alas, organic chemical nomenclature, or rather chemical nomenclature, will have to go the same way. The directed nomenclature is destined to get an ever stronger grip on chemists. Fortunately, every citizen of a democracy has the right to say that he does not admire directed economy or does not agree with certain of its measures. Of course, in the same way, chemists will be gladly allowed to declare themselves opponents of directed nomenclature, or to state that they do not approve of certain nomenclature rules, but they will be required to observe what has been internationally accepted as good nomenclature practice.

Ciphering Systems

The present internationally accepted rules in the field of systematic nomenclature do not make it possible to give every organic compound a reasonable name. Of course, this defect is felt most in the registration of literature, for example, in the preparation of handbooks and abstract journals, and in indexing such works. Obviously one can often resort to structural formulas in such cases, but these formulas are not at all suitable for systematic registration. Especially in the circles of those who occupy themselves with or are interested in documentation work, the need was felt to find a solution for this problem. Thus the so-called ciphering systems originated, and, at present, several of these are being studied by the Commission for Codification, Ciphering and Punched Card Techniques, instituted a few years ago by the International Union of Pure and Applied Chemistry.

A cipher is a series of letters, figures, and other symbols, formed according to a definite set of rules, which represents the structure of a given compound in as much detail as possible and in a unique way, and which, with the aid of these rules, is easily transformed back to the structural formula of the compound in question. Ciphering systems which at the present moment are available will not be discussed here, but these systems satisfy, to a widely different degree, the numerous requirements which can and must be brought forward. In this respect, remarkable and valuable achievements have already been reached. At its meeting in New York in 1951, the above-mentioned Commission for Codification, Ciphering and Punched Card Techniques concluded that the Dyson cipher (*3*) should be adopted as the provisional international system, and also that particular attention should be given to the inclusion in this system, where possible, of desirable features of other ciphering systems, especially those of Gruber and Wiswesser. There is little doubt that before long there will be a ciphering system which, at least from an organic chemical point of view, is entirely satisfactory.

It is certainly right that the problem of the introduction of a suitable ciphering system was assigned to a separate international commission. Indeed, special problems arise, which have little or nothing to do with chemical nomenclature—for example, problems like the use of the cipher for classification purposes, or the possibility of using it in connection with punched cards and mechanical sorting. It will, however, be necessary for this commission to cooperate very closely with the three nomenclature commissions of the International Union of Pure and Applied Chemistry. This statement is a consequence of the opinion that the aim must be to develop and to get universally accepted a cipher which can readily be transformed into a clear and easily pronounceable name—a name which is in harmony with current nomenclature. Not all of the ciphering systems so far proposed come up to this very important requirement, but without doubt very interesting possibilities in this respect are available.

The opinion has been expressed that such a cipher might replace the present organic chemical nomenclature. The correctness of this view must be denied. Even in the inconceivable case that a general readiness to take this step did exist, it would not be diplomatic to proceed with it other than very gradually. There exists an enormous amount of literature which is based on the nomenclature presently used and this literature will obviously retain its value for a long time. In the future it will be necessary to be informed of the now existing and old nomenclature practices in order to be able to use the older literature. Under such conditions it is certainly much better to improve and extend the existing nomenclature, and possibly to enrich it with some entirely new conception like a cipher, than to make revolutionary changes.

Perhaps, the primary aim might be to have the cipher, and possibly—but this is less important—the name resulting from this cipher, at least once mentioned with each compound dealt with in papers, handbooks, and abstract journals, unless the compounds are of a very simple nature. A consequence of this would be that indexes of these ciphers, and possibly of the names derived from them, would have to be added to the current indexes. Much might be gained in this relatively simple way, provided that a good ciphering system is available and that those concerned are really prepared to cooperate. The usefulness of works like Beilstein's "Handbuch der organischen Chemie," Elsevier's "Encyclopedia of Organic Chemistry," or *Chemical Abstracts* would be enormously increased in this way.

Nomenclature of Ring Systems

A particularly difficult subject is the nomenclature of the so-called parent ring systems, consisting exclusively of carbon atoms. Once a good arrangement is made in this field, the nomenclature of the heterocyclic systems need not form a serious problem. Indeed, the latter can easily be based on that of the homocyclic, nearly always carbocyclic, systems by using the "a" nomenclature, sometimes called the oxa-aza convention, proposed for this purpose independently and at about the same time by Stelzner (*14*) and by Patterson (*10*). According to this device piperidine, for example, can be indicated as azacyclohexane. The use of the "a" nomenclature for this purpose has already been sanctioned in rule 16 of the Definitive Report. A considerably improved and extended text for this rule is at present being studied by the Commission on the Nomenclature of Organic Chemistry. There is no doubt that the "a" nomenclature is going to play a very important role in organic chemical nomenclature practice.

Wisely, the Definitive Report is practically silent about the nomenclature of the homocyclic systems, unless these are monocyclic and thus do not really offer problems. Since this report was published, the International Commission has always hesitated to tackle this subject. It was discussed briefly during its meeting in Rome in 1938; however, the minutes of this meeting, published in the "Comptes Rendus de la 13e Conférence de l'Union internationale de Chimie" (*5*), do not shed much light on the problem. Fortunately, at the present moment, the situation is quite different. The International Commission has placed the nomenclature and the closely connected numbering of polycyclic systems on its program, and only quite recently discussions on this topic were started. The Commission need not complain of a shortage of documentary material in this field. Various systems for naming and numbering homocyclic systems have been proposed in the course of years; most of these—only the generally known systems of Stelzner and Patterson, and those of Dupont and Locquin (*2*) and Taylor (*15*) may be mentioned here— have already been published. In this connection the remarkable and most useful book by Patterson and Capell, entitled "The Ring Index," must be mentioned (*12*). The existence of all these systems and the fact that the leading handbooks, periodicals, and abstract journals have taken their (divergent) choices from these for their own use, may make it rather difficult to come to an international agreement. Perhaps an entirely new idea, possibly connected with the forthcoming ciphering system, might bring the best acceptable solution; such an idea is in the hands of the International Commission.

"A" Nomenclature for Open Chains

Organic chemistry is constantly growing, and obviously its nomenclature must be adapted to this development as well as possible. In consequence, the International Commission has had to occupy itself with a number of subjects completely outside the Definitive Report. The very satisfactory nomenclature of the organosilicon compounds, which was definitely adopted about 2 years ago with the energetic cooperation of American specialists in this field must be mentioned. Rules concerning so-called "extra" or better "indicated" hydrogen atoms and concerning the application of the "a" nomenclature to open chains are in an advanced stage of preparation. These latter rules will give an international basis to an idea which was expressed by various chemists independently, and which will make it possible to give simple and short names to compounds containing open chains of two or more kinds of atoms; the compound, $CH_3.O.CH_2.O.CH_2.NH.CH_2.CH_3$, to take a simple example, can according to the rules in question be called 2,4-dioxa-6-azaoctane. Without the aid of the rules now under consideration compounds with hetero chains of more or less complicated structure can hardly be named, and certainly not without resorting to the use of a large number of brackets of different kinds. The use of the "a" nomenclature will lead to a great simplification of naming. The only difficulty is to make it clear when this new nomenclature may be applied; obviously it is not the intention to have a name like dimethylamine replaced by 2-azapropane.

Cooperation with the Biochemists

The International Union of Pure and Applied Chemistry, besides appointing the Commission for the Reform of the Nomenclature of Organic Chemistry also created a Commission for the Reform of the Nomenclature of Biochemistry. The latter commission worked quite independently for several years under the chairmanship of Professor Bertrand and compiled rules for the naming of various groups of organic compounds, for which the names glucides, lipides, and protides were coined. The rules in question and the definitions of the terms just mentioned are unsatisfactory from many points of view.

Fortunately, the situation at the moment is quite different. The present international commissions for organic chemical and for biochemical nomenclature are cooperating closely. This is essential, as the problem of naming organic compounds which for one reason or another are of biochemical importance often arises, and both groups of chemists are interested in good names for these compounds. This cooperation has already led to drawing up satisfactory rules for the nomenclature of amino acids and of carotenoids and a start has been made with the study of the nomenclature of the carbohydrates—a very unhappy but generally used name—and of the steroids.

Of course, there are similar contacts with the International Commission for the Nomenclature of Inorganic Chemistry. Mutual interest is less strong here, but may possibly become stronger in the near future.

Conclusion

In the "Allgemeines Deutsches Kommersbuch" there is a song entitled "Einleitung in die organische Chemie" which is sung to the well-known tune of "Es braust ein Ruf wie Donnerhall" and which ridicules the trivial names of organic chemistry. Since this work of

>Dinitrobromanthrachison
>Alphaphenylacrosason,
>Benzol, Toluol, Xylol, Naphthol,
>Phloroglucin, Guajakol,
>Propylamin, Butylamin (bis)
>Kaliumisophthalat, Trichlorhydrin (bis)

art was written, much has changed in nomenclature, and not only in the sense that the present names are less suitable for putting to music. The future will probably bring even greater changes in the naming of organic compounds. However, it is out of the question that these names will ever lose the cacophonic character they have in the opinion of the layman. If a good name, besides meeting the many requirements which must reasonably be brought forward from the point of view of nomenclature, would also have to be harmonious to the layman, I would immediately and gladly resign my membership of the International Commission on the Nomenclature of Organic chemistry.

Literature Cited

(1) *Chemistry & Industry*, **1951**, SN 1-10, June 23.
(2) Dupont, G., and Locquin, R., *Ann. Chim.*, (12) **1**, 45 (1946).

(3) Dyson, G. M., "A New Notation and Enumeration System for Organic Compounds," 2nd ed., New York, Longmans, Green & Co., 1949.
(4) Grignard, V., "Traité de Chimie Organique," Vol. I, 1935, pp. 1073–108.
(5) International Union of Chemistry, "Comptes Rendus de la 13e Conférence," Rome, 1938, p. 36.
(6) International Union of Pure and Applied Chemistry, "Comptes Rendus de la 15e Conférence," Amsterdam, 1949, pp. 127–32.
(7) *Ibid.*, pp. 132–85.
(8) "Joint Symposium on the Nomenclature of Hydrocarbons," papers presented before the Divisions of Petroleum Chemistry, Chemical Education, Chemical Literature, and Organic Chemistry at the 116th Meeting of the AMERICAN CHEMICAL SOCIETY, Atlantic City, N. J.
(9) *La Nomenclatura Chimica*, discontinued with Vol. 5, No. 7 (July 1935).
(10) Patterson, A. M., *J. Am. Chem. Soc.*, **50**, 3074 (1928).
(11) *Ibid.*, **55**, 3905 (1933).
(12) Patterson, A. M., and Capell, L. T., "The Ring Index, Ring Systems Used in Organic Chemistry," AM. CHEM. Soc. Monograph No. 84, New York, Reinhold Publishing Corp., 1940.
(13) Pictet, A., *Arch. sci. phys. nat.*, (3) **27**, 485 (1892).
(14) Stelzner, R., "Literatur-Register der organischen Chemie," 1926. Introduction to Part V.
(15) Taylor, F. L., *Ind. Eng. Chem.*, **40**, 734 (1948).

RECEIVED August 23, 1951.

Work of Commission on Nomenclature of Biological Chemistry

J. E. COURTOIS

Faculty of Pharmacy, Paris, France

The Commission on the Nomenclature of Biological Chemistry, created in 1921 in Brussels, has met at all subsequent conferences of the Union. It has always worked in close association with other commissions of the Union, especially that on organic nomenclature. There has been progressive improvement in its method of work. At present proposals are put before the council only after long preparatory work, during which the commission consults the different national committees on nomenclature as well as the principal scientists concerned with the compounds under study. The results obtained in the nomenclature of carbohydrates, fats, proteins, and enzymes are reviewed. Since the end of the war, under the auspices of Karrer, rules have been adopted for the nomenclature of carotenoids and amino acids, the latter on the basis of reports presented by national committees of Great Britain and the United States. On the agenda of the XVIth conference are two rules on amino acids and the nomenclature of steroids and vitamins. Many rulings of the commission have become established in practice. Others have not found universal assent, but even in these cases the commission has helped to reduce confusion.

Biochemical research on a world-wide scale began in the years 1905 to 1910. About 1920 biochemists realized that it was becoming essential to coordinate the nomenclature of natural compounds.

The Nomenclature Commission

The International Union of Chemistry, founded in 1919, had as one of its first achievements the instituting of an International Commission on the Nomenclature of Biological Chemistry. At the second conference of the Union, held in Brussels in June 1921, R. Marquis, professor at the Sorbonne, presented a report justifying the establishment of this new organization to study nomenclature.

The Union then created three international commissions on nomenclature on: inorganic chemistry, organic chemistry, and biological chemistry. Each of these commissions contained a delegate from each of the national organizations of chemistry represented in the Union. A smaller working party was planned in the form of a committee of six members. The first members of this committee, who probably represent the first international group of biochemists ever to be created, were: G. Bertrand of France, Sir Arthur Harden of Great Britain, P. A. Levene of the United States, A. Pictet of Switzerland, and

S. P. L. Sörensen of Denmark. The sixth place was reserved for a representative of the Slavonic languages who was to be nominated later.

The first task of the working committee was to constitute national committees on nomenclature under the auspices of the national chemistry organizations. These national committees were to draw up reports which would subsequently be communicated to the working committee. Nations not represented on the working committee were invited to send their proposals to every member of the working committee.

To simplify this exchange of documents, the president and the secretary of the nomenclature commission later undertook this work of coordination and distribution. They also maintained contact with the board of the International Union of Chemistry through its permanent secretarial body.

The International Commission on the Nomenclature of Biological Chemistry began to function immediately, meeting for the first time in Lyon, France, in June 1922. As the basis for its work the commission decided upon a very thorough report by Bertrand (1), a distinguished biochemist who was to be chairman of the commission for many years.

Table I shows the commission's activity during its 30 years' existence. The vigor and activity of the commission are demonstrated by the fact that meetings took place regularly at each conference of the Union.

Table I. Conferences of International Union of Chemistry

Conference	Year	Place of Meeting	Chairman Presiding over Discussions	Subjects Treated	No. of Members Present	No. of Countries Directly Represented
III	1922	Lyon	E. Paternò	Preliminary work	15	11
IV	1924	Cambridge	G. Bertrand	Bases for classification of principal biochemical constituents	14	7
V	1924	Copenhagen	G. Bertrand	Carbohydrates (glucides)	19	10
VI	1925	Bucharest	E. Votoček	Carbohydrates (glucides)	17	10
VII	1926	Washington	G. Bertrand	Proteins	15	6
VIII	1927	Warsaw	G. Bertrand	Glucides Proteins Enzymes	16	10
IX	1928	The Hague	G. Bertrand	Fats	13	8
X	1930	Liége	G. Bertrand	Future organization of work	Not mentioned in report	Not mentioned in report
XI	1934	Madrid	G. Bertrand	Modifications of methods of work	18	10
XII	1936	Lucerne	Sir Arthur Harden	Enzymes	4	4
XIII	1938	Rome	G. Barger	Enzymes	5	5
XIV	1947	London	P. Karrer	Carotenoids Amino acids	6	5
XV	1949	Amsterdam	P. Karrer	Amino acids Vitamins	6	5

In the early stages there were few national bodies for studying problems of nomenclature. Therefore it was necessary for the commission to call upon individual scientists able to represent the opinion of the biochemists of their country. The procedure was also established of having several representatives from the country where the conference was being held participate in the discussion: five Danes at Copenhagen, three Romanians at Bucharest, four Americans at Washington, and three Poles at Warsaw.

At early meetings, members of other commissions, such as that on organic chemistry, were also present. The list of members at the different meetings reveals a certain diversity, that cannot be avoided when an organization begins to function, but it was one of the reasons why the first proposals of the commission did not always prove acceptable to all countries. It was found necessary therefore to modify the procedure. This was effected between 1930 and 1933.

At this period many national nomenclature committees were at work in different countries. The commission felt that above all its role should be further centralization and coordination of the first work done—it could make proposals only when suggestions had been submitted to it; and it could make decisions after submitting its projects for examination to the different national bodies affiliated with the Union (6).

As such decisions could be made only with the sanction of the Council of the Union, it became necessary for the Commission on the Nomenclature of Biological Chemistry to work in contact with other commissions of the Union.

This has been effective for 20 years with the Nomenclature Commission of Organic Chemistry. A. M. Patterson, organizer of the symposium on Chemical Nomenclature, and P. E. Verkade, chairman of the Nomenclature Commission of Organic Chemistry, have contributed much to the smooth working of these two commissions. There is no fixed procedure governing relations between these two bodies. There is just one rule which has always been followed—to work in close contact for a common purpose. Between conferences the chairmen and secretaries of the two commissions communicate with each other and exchange reports of common interest. During conferences, questions related to biochemistry but of interest in the nomenclature of organic chemistry are generally first examined at the meetings of the biological chemistry commission.

Subjects selected by this commission for proposals are examined at joint meetings of the organic and biological chemistry commissions; they are then put forward at the Council of the Union. When approved by the latter, proposals are sent for the last time for consideration by the competent national committees, and the final decision is made by the Council at its next conference.

Table I shows that since 1936 fewer members have been present at sessions of the commission. This implies no lack of interest, but is a result of changed methods of work. The commission has become a centralizing body, a group of from 5 to 10 biochemists representing as many different nations and languages as possible. Commission members are usually recognized specialists with regard to questions under consideration. At the Amsterdam conference in 1949, when it was decided to start upon the nomenclature of vitamins and steroids, the commission requested the Council of the International Union of Chemistry to affiliate new members specializing in these problems. As with other commissions of the Union, the new members put forward after voting by the commission are then nominated by the Council and the executive committee of the Union.

Current Commission. The International Commission on the Nomenclature of Biological Chemistry is at present composed of: chairman, J. Murray-Luck, Stanford University, Calif.; E. Cherbuliez, Geneva, Switzerland; J. E. Courtois, Paris, France, who acts as recording secretary; A. H. Ennor, Melbourne, Australia; Sir Charles Harington, London, England; B. C. P. Jansen, Amsterdam, Holland; G. F. Marrian, Edinburgh, Scotland; Byron Riegel, Evanston, Ill.; and A. Rossi-Fanelli, Rome, Italy.

Work Methods. When the commission decides to examine a problem, it is frequently at the request of a national institution of nomenclature. In the course of this survey not only are groups affiliated to the Union consulted, but also distinguished scientists who are specialists on the subject. New substances extracted from a biological product usually keep the name given by the person who made the discovery. This is possible only when the name is not too remote from the usual rules of nomenclature.

It is preferable that the name should be related either to the original substance or to a characteristic chemical property. Names associated with physiological properties that may imply medical use are rightly not accepted by the editorial boards of many periodicals. This is particularly the case with medical or pharmaceutical journals in the United States.

When the name originally suggested by the person who made the discovery is to be replaced or modified, the commission approaches the scientist himself.

Thus when proposals were being considered for the nomenclature of the amino acids, Rose (*12*) agreed to modify the term "$d\,(-)$ threonine" he had suggested to show the connection with D-threose to L-threonine, in conformity with the rulings proposed for the other amino acids.

Preliminary consultations must be on a broad scale, so that the commission may account for all existing tendencies. Nomenclature of enzymes is an example of this preparatory survey. At Lucerne in 1936 (*8*), the commission drew up a preliminary report which was submitted to almost all recognized enzymologists. Although this work was interrupted by the war, the following details were obtained from the records of the commission.

P. Karrer communicated the commission's proposals to 23 German or Scandinavian scientists. He received 17 replies: 13 in agreement with all the rulings, 4 suggesting modifications.

Sir Arthur Harden sent eight letters from British biochemists proposing more or less important modifications.

R. A. Gortner, University of Minnesota, was at that time chairman of the American Committee on Biochemical Nomenclature. He received 11 letters from American enzymologists with modifications to the original text.

René Fabre, for 20 years the commission's efficient and devoted recording secretary, contacted biochemists of the Latin and Eastern European countries.

The commission's proposals were submitted to 68 enzymologists of 14 different nations. Thirty replies, some of which were most detailed, had been received when war broke out and stopped the survey.

This was the indispensable method of work the commission was to utilize when functioning after 1945.

Progressively developed in the course of 30 years' work, the methods of the commission should justify hopes that its carefully considered proposals will meet with general agreement.

Necessity for Coherent Nomenclature and First Work of Commission

Bertrand's general report (1), adopted as the basis for the commission's work, clearly defined the need for coherent nomenclature. Some of his apt remarks are still remarkably true:

The number of definite organic substances extracted from plants or animals reaches a considerable figure and is continually being increased. Each substance when discovered must be given a name.

In most cases the choice of names was determined by no general ruling and was simply the product of the scientists' fancy. The names most happily chosen were related either to the botanical or zoological origin of these substances or to one of their physical, chemical, or physiological properties. One rarely finds a particular termination giving some exact indication as to the position they occupy in the series of compounds of organic chemistry.

Thus the list of natural compounds abounds in a variety of names which are often meaningless from a chemical point of view and may be very different for substances closely related to each other, or similar and even identical for different substances.

The decision of the Nomenclature Commission for Biochemistry logically therefore was to declare in 1923, as its first rule, that "The name of a natural substance, the chemical constitution of which is known, must be formed in accordance with the rules of the nomenclature of organic chemistry." It was, of course, useless to expect that this rule would be punctiliously observed by all, but on the whole it was observed. By comparing the names selected from the beginnings of biochemistry up to 1920 with the even more numerous ones proposed since then, there is clearly observed greater coherence in the latter.

The primary aim of a nomenclature commission should be to limit disorder and to introduce logical order. It is from this point of view that the commission's first decision has borne fruit.

This first general decision was completed the following year with two other rules which were satisfactorily observed, despite numerous exceptions:

In the event of the constitution of a natural substance being too complex or not well known, the name with which it is to be designated should, at any rate, feature a termination in agreement with the principal chemical group.

The termination -ine shall henceforth be used only for natural substances containing nitrogen in an alkaline group, with the possibility, in each country, of using the ending -in or -ine.

It was on the basis of these rules that the commission began work on the classification of the principal groups of fundamental biochemical compounds: carbohydrates, lipides (fats), and proteins.

Carbohydrates (Glucides)

Until 1920, the terms "carbohydrates," "*Kohlenhydrate*," and "*hydrates de carbone*" did not have exactly the same meaning in the different countries. For some they applied only to reducing sugars, for others they also covered the polysaccharides and glucosides. Even for simple sugars (monosaccharides) the term "carbohydrate" was not logically acceptable. It applies in theory to substances with a rough general formula $C_nH_{2m}O_m$ — to substances containing hydrogen and oxygen in the same proportion as in water. It is not applicable to a certain number of very common chemical compounds: methylpentoses, desoxyribose, aminated sugars, etc.

The commission could not find a satisfactory definition for all with the term "carbohydrate." After long discussion it finally adopted the term "glucide" proposed by Bertrand.

At Cambridge in 1923, the commission decided by a vote of 10 out of a total of 14 votes to carry the following motion:

The term "glucide" designates the group of substances comprising reducing sugars and compounds giving by hydrolysis one or more of these reducing sugars.

Experience showed, however, that it was very difficult to invent new terms that would be universally accepted. The term "glucide" has been regularly used in all French-speaking countries, but it is rarely used in chemical literature in the Anglo-Saxon and German languages. But, subsequent to the indirect influence of the commission's decisions, the term "carbohydrates" has in practice become synonymous with "glucides" from the point of view of definition.

The commission then turned to nomenclature for the constituents of the glucides, subdividing them into nonhydrolyzable reducing sugars (monosaccharides) and into substances split by hydrolysis into one or more reducing sugars, accompanied or not by other substances.

Simple Reducing Sugars (Monosaccharides). The term "monosaccharides" did not appear satisfactory for several reasons. In the first place, it had the disadvantage of applying to a whole group of substances, with habitual names ending in -ose. In addition, it was not perfectly logical and created confusion with the saccharose formed by the association of two monosaccharides which thereby became a disaccharide. So the commission set about selecting a term with an -ose termination like all constituents of the group.

Bertrand suggested "glucose," the most typical example of the group. This denomination, adopted by the commission in 1924 at Copenhagen, was somewhat contrary to former practices and aroused criticism.

In 1927, Bertrand (*2*) presented a proposal by H. Hérissey and M. Bridel (*3*) suggesting that only the terminations -ose and -oside be retained of the words "glucose" and "glucosides" used as generic terms. The Warsaw conference adopted this designation of oses for simple sugars. The contraction was a simple one and made it possible to return easily to such customary nomenclature as aldoses, ketoses, pentoses, and hexoses.

Substances Yielding Reducing Sugars When Hydrolyzed. From analogy with the definition of the oses, it was decided to call these substances "osides." They were subdivided into two categories: "holosides" yielding only oses by complete hydrolysis and "heterosides" yielding by hydrolysis one or several oses accompanied by nonglucide substances (aglycons).

The commission proposed to represent the classification of glucides as shown in the following diagram:

```
                 oses
                /
       Glucides      holosides
                \   /
                osides
                    \
                    hétérosides
```

Teaching of Biochemistry. French biochemists unanimously adopted the propositions of the commission in their teaching. Such conformity may seem surprising on the part of representatives of a nation renowned, rightly or wrongly, for its inclination toward indiscipline and its cult of individual freedom. But it is certain that in 1927 all eminent French biochemists had assisted in fixing the nomenclature advocated by Bertrand and Bridel at the international conference, so it is not surprising that this nomenclature of the glucides should be tried out in practice by those who had proposed it. The trial was conclusive. For more than 20 years French biochemists have used in their teaching the rulings established by the commission for the nomenclature of glucides. These rulings are simple, clear, and logical. The Cartesian desire for precision easily triumphed over an inclination toward indiscipline.

In other countries, where the new word suggested could not meet with success, the subdivision of glucides into their principal groups was, however, adopted.

Textbooks of biochemistry in different languages now classify the constituents of the glucide group into subdivisions corresponding to those in oses, holosides, and heterosides. It can now be assumed that in teaching the decisions of the commission are respected in spirit, if not in letter.

Papers of Original Research. The decisions of the commission have made it possible to create some order, even when they were not entirely respected, and they have limited disorder.

The term ose is certainly the one most subject to criticism even though it proves satisfactory from the point of view of termination with the usual names of all simple sugars ending in -ose.

When the commission began working, this termination had already overstepped the bounds of simple sugars and was being applied to reducing or nonreducing oligoholosides. Even French authors most respectful of the rulings on the nomenclature of glucides continued to use the ending -ose for these holosides. They found it impossible to use neologisms such as saccharoside for saccharose, and stachyoside for stachyose.

Similarly, the ending -oside has rarely been applied to the polyholosides. Inulin and starch have logically retained their former names.

Votoček emphasized this partial failure in the report he presented in 1947 at the London conference (15). Votoček proposed for simple nonhydrolyzable sugars the generic term "protoses." This denomination did not meet with general approval.

The -oside termination is only very rarely used for the holosides, but it is being regularly adopted for the new heterosides. In various countries the heterosides that have long been known have retained the termination -ine, although not containing amino nitrogen (salicine, digitalin). It is regrettable that, owing to the persistence of old habits, the name glucoside has not been restricted solely to heterosides yielding glucose when hydrolyzed. Appellations in which the name of the sugar precedes the ending -oside are far more logical and precise, such as galactoside, rhamnoside, and arabinoside.

When the commission turned to the question of the glucosides, one of its main purposes was to bring some order to what was to become the group of heterosides. On this point, its efforts, though rarely appreciated, have none the less proved fruitful.

Lipides

As early as 1924 the commission, which had adopted the neologism "glucide" with only a majority vote, decided unanimously to change the name "lipoid" to "lipide"—indicating fats and esters possessing analogous properties.

In 1928 the commission agreed to continued use of the term "lipoid," but recommended that it should not be employed as a noun in the chemical sense, but as an adjective with a physical meaning; in such a case it would be preferable to replace it by the word lipoidic.

With greater homogeneity in mind the commission decided that the names given to lipides shall terminate in the ending -ide. This rule was not universally adopted; in many countries the denominations olein, palmitin, lecithin, and cephalin continue to be used. It is not regularly applied even in France.

At the Hague in 1928, the commission was concerned with fixing the terminology of the chief constituents of the lipide group. It divided them first into ternary lipides and conjugate lipides.

The ternary lipides are those containing only carbon, hydrogen, and oxygen but not phosphorus and nitrogen. They are divided into:

Glycerides, in which the alcohol is glycerol
Cerides, lipides formed by the union of higher monovalent alcohols and fatty acids with a generally high molecular weight
Sterides, in which the alcohol is a steroid
Etholides, lipides formed by hydroxy acids where the acid group of a molecule forms an ester with the hydroxyl of another molecule.

For the conjugate lipides the commission proposed that the term "phosphatides" be replaced by "phospholipides" and "phosphoaminolipides," defined as:

Phospholipides, lipides containing phosphorus (in the form of phosphoric acid residues);
Phosphoaminolipides, lipides containing both phosphorus (in the form of phosphoric acid residues) and nitrogen (in the form of amino residues). These phosphoaminolipides were themselves subdivided into:
Glycerophosphoaminolipides, in which the alcohol is glycerol (cephalins, lecithins) and
Sphyngophosphoaminolipides, in which the alcohol is sphingosine (sphingomyeline). Cerebrosides not possessing an ester linkage are not included in the lipide group but in that of the heterosides.

These decisions did not meet with the approval of the Anglo-Saxon representatives at the 1930 conference in Liége. It was decided, however, that a report by Vesely and Jakes (*13*) be taken as a basis for discussion at the following conference, after being first submitted to the different national committees. No decision has been taken on this subject since 1930.

The 1928 suggestions were criticized principally for being too vague on certain points and perhaps overhastily detailed in other cases.

Vesely and Jakes (*13*) pointed out that the definition for lipides adopted in 1923 was not sufficiently precise. It included, among lipides, the fats formed of mixtures of esters and nonlipidic substances (free fatty acids, steroids, carotenoids, hydrocarbons); on the other hand, certain essential oils formed mainly of esters might be considered lipides. Vesely and Jakes advocated a more precise and concise definition for lipides: "natural esters nonvolatile with steam and possessing no aromatic ring in their molecule."

The definition for cerides was also subject to criticism for, if applied literally, it included steroids in the groups. Vesely and Jakes (*13*) proposed that cerides be fixed as "lipides formed by the union of aliphatic alcohols and fatty acids with a generally high molecular weight."

Proposals regarding the nomenclature of ternary lipides were very detailed and therefore more difficult to adopt.

The practice has been established of naming the different phosphoaminolipides "phospholipides." But what the commission proposed to name phospholipides are most currently termed phosphatidic acids.

If the classification of lipides were to be examined, some points would have to be made clear. In the first place, a definition of lipides should be made that excludes esters of mineral acids such as orthophosphoric acid. The phosphoric esters of alcohols and phenols are alkalino-stable—excluding those possessing a free carbonyl group; moreover, they are not very soluble in organic solvents. The definition of lipides should therefore take into account the hydrolysis of ester linkage by heating with alkalies and the solubility of lipides in some organic solvents.

It should also be useful to find a definition rendering it possible to include among the lipides a group of compounds that are now usually placed there—the acetal-phosphatides—formed from two alcoholic groups of glycerol and the aldehydes of the corresponding fatty acids.

The classification proposed by the commission for the lipides, a most heterogeneous group, was without doubt less logical than the one advocated for the glucides. That may explain its smaller success. It seems, however, that the proposals advocated by the commission in 1928 might still serve as a basis for discussion in the attempt to fix rules adapted to what has become the most general practice.

Protides

The nomenclature of this group, the most important and the most complex compounds of biochemistry, was begun in 1923. Bertrand (1) suggested a classification based on the same pattern as that of the glucides—the amino acids being considered as characteristic of the protides as the reducing sugars were characteristic of the glucides.

In Cambridge in 1924 the commission decided unanimously that the term "protide" should apply to the group of substances which comprises the natural amino acids, and should also apply to those substances which yield upon hydrolysis one or several of these amino acids.

This definition should be corrected by eliminating the word "one," for it is impossible to consider as protides a large number of compounds which liberate upon hydrolysis one molecule of amino acid, such as glycocholic acid which yields glycine and cholic acid; pantothenic acid which liberates β-alanine; and phospholipides containing serine. At the most, hippuric acid can be considered as a natural, somewhat special peptide.

In 1926 in Washington the commission considered a temporary classification of protides into two groups: amino acids, and those compounds which liberate upon hydrolysis these amino acids, whether or not they are associated with other substances. In Warsaw in 1926 this second group was subdivided into peptides, holoproteides, and heteroproteides.

Peptides result from the union of several amino acids, where these substances are united by linkages resulting from the loss of one molecule of water between the amino group of the first molecule of amino acid and the carboxyl group of the next one. The term "peptide" previously suggested by Emil Fischer has received general approval. Certain authors have made this classification even more complete by distinguishing between the oligopeptides having a small number of amino acids and the polypeptides having a larger molecule.

Holoproteides (holoproteins) liberate upon hydrolysis only amino acids or ammonia. This definition should be interpreted according to its spirit and not to the letter, as several purified albumins and globulins yield upon hydrolysis hexoses, osamines (aminosaccharides), and phosphoric acid. Holoproteides have been divided into a certain number of groups, according to the Anglo-American proposals: protamines, histones (or histonines), albumins, globulins, glutenins (glutelins), gliadins, scleroproteins, and keratins. All these terms have been in general use ever since, whereas the term holoproteide (holoprotein), which was a neologism, has met with less success. The commission did not attempt to specify the characteristic features of the subgroups of the holoproteins. It is rather difficult to determine with accuracy their characteristic traits: solubility, salting out, composition, and behavior in the presence of proteases. For example, the plant globulins can be grouped together with the other animal globulins in regard to their solubility in saline solutions and their electrophoretic mobility; on the other hand, these two groups of globulins behave differently when being salted out.

Heteroproteides (heteroproteins) liberate upon hydrolysis amino acids and ammonia together with other nonprotein substances. The commission suggested their subdivision into such groups as chromoproteins and phosphoproteins. This classification has, in general, been adopted. In common use in several countries, "heteroproteide" has substantially the same meaning as the terms "*zusammengesetze Proteine*" or "conjugated proteins," used in other countries.

Even though the distinction between holoproteides and heteroproteides can give rise to some criticism, nevertheless it is of unquestionable didactic value. To the students studying biochemistry, it marks clearly the difference which separates the two groups—the difference which is manifest in the main chemical properties as well as in the part which they play in the intermediate metabolism.

General Remarks on These Classifications

An examination in their entirety of the results of the work of the commission to establish the nomenclature of the principal groups of biochemical compounds can be summarized:

It is difficult to get all biochemists to accept the new names suggested for compounds or groups of compounds.

The compounds described after the establishment of the commission's rules of classification have, in most instances, been named in conformity with these rules.

The classifications suggested by the commission have, as a whole, been adopted everywhere.

As an indirect result of the work of the commission, the names which have been kept for certain groups of compounds have, in practice, acquired the same significance as the new terms which were suggested: *Kohlenhydrate* or carbohydrates and glucides; *Eiweiβ-stoffe* or proteins, and proteides. And even though the decisions of the commission have not been applied to the letter in all countries, at least they have, in most instances, been adopted in their spirit.

Nomenclature of Limited Groups of Biochemical Compounds

Starting in 1935, the work of the commission has had for its primary object the study of more limited groups of biochemical compounds: enzymes, carotenoids (carotinoids), amino acids, vitamins, and steroids.

Enzymes. In 1877, E. Duclaux published the first general survey on enzymology (7). Duclaux suggested that the suffix -ase should be attached to all the names given to enzymes, in order to show their relation with the first enzyme whose existence had been clearly demonstrated—the diastase from germinated barley, isolated from the malt by Payen and Persoz in 1833.

At the Warsaw conference in 1927, the commission adopted the following rule:

The term ases is suggested in order to designate the totality of the soluble ferments (diastases or enzymes). The names of all the ases must have the suffix -ase.

The neologism ases has not met with great success. Even the commission did not refer to it in later official reports. On the other hand, the recommendation to use the suffix for all the enzymes has probably been the most closely adhered-to decision. The enzymes discovered since 1927 which do not possess this suffix are very scarce indeed.

In 1936 in Lucerne the commission decided to study the nomenclature of the enzymes. This decision really marked the official approval of this last term, as compared to its synonyms. The term "enzyme" was introduced at the suggestion of A. Harden (10). The old name "soluble ferment," as opposed to "figured ferment," no longer corresponded to the progress made in the field of enzymology. The development of the research on intermediate metabolism had already shown the fragility of this distinction. At the most, a vague similarity could be found with the very practical mode of expression suggested by Willstätter, who distinguishes between the lyoenzymes, directly extractable, and the desmoenzymes, more solidly bound to the tissues. The term "diastase" is not being used any more. French authors, however, have preserved a certain sentimental attachment toward this term and books on enzymology published in France are often entitled "The Diastases." This title marks the relationship which exists between French enzymology and the work of E. Duclaux, Pasteur's most brilliant pupil (7).

The term "biocatalyst" was, on the other hand, too general to be applied to the enzymes alone.

In 1936, the commission started to elaborate the precise terms for the nomenclature of enzymes. World War II prevented the commission from finishing its work. A great number of answers received by the commission have been compiled by its former secretary, René Fabre, who kindly put his files at the author's disposal.

The main results of this inquiry reflect the state of the problem ten years ago, and, with a few modifications and improvements, they could serve as a basis for the resumption of the work on this subject by the commission.

The main points submitted to the inquiry were:

The word to which the ending -ase has been added should preferably indicate the nature of the attached substrate (example, peptidase), or the mode of action of the enzyme (example, dehydrogenase), or a combination of the name of the substrate and the mode of action each time that it is necessary to avoid an ambiguity (example, succinyldehydrogenase).

All the answers have been favorable, with the exception of a few reservations of minor importance. Custom has sanctioned this rule, which has a rather large field of application.

The question, whether it is advisable to use special terms in order to indicate the synthetic or analytic action of an enzyme, has been left open.

At the beginning of its work, the commission had considered the possibility of using the suffix -ese for the synthetizing enzymes. As early as 1936, many answers were unfavorable to this possibility. Since then many enzymes catalyzing the reaction of equilibrium have been shown to exist. It seems that, if the question were submitted today, almost all the answers would confirm the established custom of using the suffix -ase in the names of enzymes, regardless of the state of equilibrium to which they lead a reaction.

The word enzyme designates the whole of the active complex, including the carrier and the activator. The coenzymes will be designated by the name of the activated enzyme, preceded by the prefix co-, for example, coglyoxylase. When it is necessary to distinguish between the enzyme as a whole and the enzyme deprived of its activator, the total complex will be called "holoenzyme," and the residue, after separation of its activator, will be called "apoenzyme."

This was the proposed rule which aroused the largest number of objections. The distinction between the apoenzyme which is thermolabile, colloidal, and almost always a protein, and the coenzyme, which is more thermostable and often noncolloidal, is of definite didactic value. In a basic course, it helps to point out the complexity of the enzyme constituents.

As far as a more precise mode of expression is concerned, the denomination "enzymatic system," suggested by many authors, is more satisfying to logic than the term "holoenzyme." This last term, however, can be used in a precise sense, in the majority of cases.

The definition of the coenzyme, although accepted in its essence, has also given rise to some objections, the most important being that many different, distinctly specific enzymes must have the same coenzyme in order to act. Many examples can be found in the group of the dehydrogenases. The determination of the structure of numerous coenzymes would take this objection into account—one needs only to remember the great classical rule of biochemical nomenclature: Denominate a substance by its precise chemical name whenever possible. The custom of designating the different enzymes, possessing the same coenzyme, by precise chemical names has become prevalent—examples, alloxazine nucleoproteins, pyridine nucleoproteins.

The diversity of the enzymatic systems makes it impossible to use a strict and inflexible terminology. That suggested in 1936 turned out to be acceptable, if used in a flexible manner. The proposals which were submitted to the inquiry should be considered rather as recommendations of a general character than as strict rules to be applied in all countries for all enzymes.

Carotenoids. A preliminary report on nomenclature of the carotenoids, drawn by the Committee on Biochemical Nomenclature of the National Research Council of the United States, was published in 1946 (*11*). P. Karrer, president of the commission at that time, with this report as a basis, presented a project to the commission (*9*) containing a few modifications and extensions of the American draft. Karrer's (*9*) text was discussed at the London conference in 1947 at a meeting held by the joint commissions for nomenclature of organic chemistry and biological chemistry. A few slight modifications were made, and this new text was adopted by the Council of the Union in 1947. It has been published by several periodicals (*9*) since that time.

A report of the Committee on Nomenclature of the National Research Council, suggesting some modifications, was examined at the September 1951 conference in the United States.

Amino Acids. The nomenclature of the amino acids, confusing until recently, was first considered by the commission in 1947, after some excellent preparatory work by two national organizations on nomenclature.

At the incentive of Chibnall (4), the editorial boards of the *Journal of the Chemical Society* of London and the *Biochemical Journal* worked out a first project, after consulting numerous biochemists of Great Britain and of other countries. This first text (14) made it possible for Sir Charles Harington to submit a project on amino acid nomenclature to the commission in July 1947 at the conference of London. At the same time, the commission considered a second project, worked out by the commission for nomenclature of the AMERICAN CHEMICAL SOCIETY and the editorial board of the *Journal of Biological Chemistry* (12).

The commission accepted the British project as a basis for discussion. These two projects had many points in common but they also contained some rather substantial differences. The most important of these differences were gradually settled between 1947 and 1949, through the continuous efforts of A. M. Patterson, H. B. Vickery, and Sir Charles Harington.

In 1949 the Amsterdam conference adopted the wording of seven of the nine suggested rules. Rather important differences existed in regard to the wording of rules 6 and 7 (5). The commission preferred to postpone the study. In March 1951, the British and American representatives reached an agreement on the presentation of a common text, which had been approved by their two national committees. This text, sent to the other national organization by the secretary of the chemical Union, was studied during the September 1951 conference.

Among the most important difficulties to be overcome before a definite agreement could be reached were:

The case of threonine, which could be considered in different ways from the viewpoint of stereochemistry by regarding it either as a derivative of a sugar, the D-threose, or on the contrary, as related to the other natural amino acids of the L series.

The choice of a name of common use for certain natural amino acids which can be considered as deriving from other amino acids having a trivial name: phenylalanine.

Thus, for example, the compound derived from proline having the hydroxyl group in position 4 has been, so far, the only hydroxy derivative of proline described among the natural proteins. It could therefore be called L-hydroxyproline, without great risk of confusion, or preferably by the more exact terminology: hydroxy-4-L-proline.

The work methods used by the commission for the determination of the nomenclature of amino acids have, certainly, been slow, but the purpose was to avoid hasty decisions which would run the risk of not being accepted. This is effective, for experience has shown that rules which had been established too rapidly had great difficulties in being accepted later on.

Vitamins. In 1949, in Amsterdam the commission studied a preliminary report by B. C. P. Jansen. It was accepted as a basis for discussion (5) and was sent for inquiry to different national organizations and to a large number of experts on terminology. The answers that have been received show that this is a very important problem, which will continue to be studied at the next conference. The subject matter is too complex for any definite decisions to be made in the near future.

Steroids. This problem is more limited than that of the vitamins. Some excellent preparatory work was accomplished at a conference held at the Ciba Foundation in London, on May 30 and June 1, 1950. This preliminary report has been revised and amended. Byron Riegel has sent it to the members of the Commissions for Nomenclature of Organic Chemistry and Biochemistry.

This report offers a remarkable basis for work, for it has already received the approval and the signature of twenty steroid chemistry experts of different nations.

Outlook for the Future in Biochemical Nomenclature

It is very difficult to foresee in what direction the future work of the Commission for Nomenclature of Biochemistry will be oriented. It can be assumed, however, that while continuing the study of the nomenclature of the vitamins and steroids, the commission will also start considering other limited groups of compounds: carbohydrates, cyclitols, and enzymes. Projects for their nomenclature have appeared in different periodicals.

It is also probable that, sooner or later, the commission will have to consider a codification of the abbreviations which are being made use of more and more frequently. This tendency to designate biochemical compounds by initials risks making some of them unintelligible. Abbreviations like "ATP" for adenosine triphosphate, and "DPN" and "TPN" for pyridine nucleotides have been sanctioned by custom and are easily understood. On the other hand, it seems not very logical to see the coenzyme of the acetylating enzymes designated by the letter A. It is very difficult to explain in a basic course that this coenzyme has been among the most recent ones to be discovered and that, nevertheless, it has the right to the first letter of the alphabet. The same holds true for the abbreviation "GP," which is being used for designating the two glycerophosphates, the glyceraldehyde phosphate, the different glucose phosphates, and sometimes even the phosphoglyceric acids. The use of initials is very convenient, but it must remain logical and as homogeneous as possible. Biochemists should consider the establishing of some rules before too many bad habits will have become sanctioned by custom.

This report shows that it is easier for a commission for nomenclature to prevent than to cure. In the future, the work of the commission should permit the rapid naming of groups of new compounds or of those whose importance is increasing. Thus at the first stage the commission could formulate suggestions, with the definite decisions being taken only afterward in accordance with the reaction to these suggestions.

Thirty years of work by the commission have shown that classifications can be accepted, but that it is difficult to introduce into common use a neologism in place of another term, illogical as that term may be, if it has been in use for a long time.

The inspection of the results that have been obtained shows that the commission must endeavor to attain a double aim: to limit the prevailing confusion and contribute toward as much order as possible. With progressive improvement in its working methods, the commission will continue to fulfill this aim.

Literature Cited

(1) Bertrand, G., *Bull. soc. chim. biol.*, **5**, 94–109 (1923).
(2) *Ibid.*, **9**, 854–6 (1927).
(3) Bridel, M., *Ibid.*, **8**, 1211–16 (1926).
(4) Chibnall, A. C., *Biochem. J.*, **41**, XXXIV (1947).
(5) Courtois, J., *Bull. soc. chim. biol.*, **31**, 1388–402 (1949).
(6) Delaby, R., *Endeavour*, **9**, No. 33, 18, 20 (1950).
(7) Duclaux, E., "Action de diastases ou ferments solubles," in "Dictionnaire des Sciences Medicales de Dechambre," Vol. 2, pp. 667–75, Paris, 1877.
(8) Fabre, R., *Bull. soc. chim. biol.*, **18**, 1727–40 (1936).
(9) Karrer, P., *Ibid.*, **30**, 150–6 (1948).
(10) Harden, A., Comptes Rendus de la Douzième Conference, IUC, Lucerne, 1936, p. 43.
(11) Murray-Luck, J., Strain, H. H., and Mattill, H. A., *Chem. Eng. News*, **24**, 1235–6 (1946).
(12) Rose, W. C., *J. Biol. Chem.*, **115**, 721–9 (1936).
(13) Vesely, M., and Jakes, M., *Bull. soc. chim. biol.*, **12**, 128–9 (1930).
(14) Vickery, H. B., *J. Biol. Chem.*, **169**, 237–45 (1947).
(15) Votoček, E., *Bull. intern. acad. tchèque. sci.*, **54**, No. 12 (1944).

RECEIVED August 1951.

Nomenclature in Industry

H. S. NUTTING

The Dow Chemical Co., Midland, Mich.

> The economic importance of chemical nomenclature to industry is discussed. A standardized nomenclature is necessary for the internal communications and the permanent records of a large industrial organization. It is important in communications with others, particularly when large numbers of compounds are involved. Efficient use of the published literature is important to industry, and this in turn is dependent upon the use of good nomenclature by authors and editors.

Nomenclature is one of the important tools of chemistry. If it is an important tool, then it must be of value to chemists.

There are some who are inclined to believe that this value is primarily esthetic in nature, or at least a matter of scholarship. But those who have had to cope with nomenclature problems in industry have come to realize that it has economic values as well.

The purpose of this paper is to describe the role which nomenclature plays in industry, and to discuss briefly the economic aspects of nomenclature.

Types of Nomenclature

A clear understanding of the different types of nomenclature which are in everyday use is important. The names which are currently used to distinguish one item or substance from another can be divided into three general types: systematic names, common names, and trade names.

As far as chemical names are concerned the systematic names disclose not only the number of atoms in a given compound but their relationship one to the other as well. There are often several systematic names which can be used to describe a given compound accurately. Since many organic compounds are quite complex, the systematic names are sometimes too complex and cumbersome to be practicable for everyday communication. This has led to various procedures for simplifying the names. It is common practice even in systematic nomenclature to designate large molecular units by a common name such as "naphthalene" and to indicate substitutions thereon in a systematic manner.

Common or trivial names are in certain respects the nicknames or the slang of the profession. These names originate in various ways: Some are assigned before the compounds are identified; some have been inherited from the past; some are abandoned or unregistered trade names; while still others are simply nicknames or abbreviations. Occasionally the common names give a clue as to the structure of the compound, but in most cases the common name is entirely nondescriptive.

Trade names should be mentioned only briefly, inasmuch as they are, for the most part, the private property of a given organization. They are therefore not available for general use except for identifying the product as it is supplied by the owner of the trade-mark.

Nomenclature Problems in Industry

Actually, the nomenclature problems of industry are not far different from those encountered in other lines of activity. And yet the attitude of the modern industrial scientist toward nomenclature is often different from that of one outside of industry. The reason is that the industrialist is trained to be a member of a closely coordinated team, and he is reminded almost daily of the importance of keeping communications as clear as possible, while others are not so accustomed to thinking in terms of teamwork.

Like anyone else in the laboratory, the industrialist often uses nicknames to designate the things he is working with day by day. Where chemicals are concerned it is common for one person to prefer one name for a given compound and another a different name. It is conceivable that in small organizations the use of nicknames—or several names for that matter—for the same material can be used by everyone without impairing appreciably the efficiency of operation.

As the size and complexity of the industrial organization increases, however, the laissez-faire approach to nomenclature tends to become less and less satisfactory. A disproportionately greater amount of time seems to be required by everyone concerned because of nomenclature problems. To make this point more specific, two practical case studies will be described.

Usually it is no particular problem to discover whether or not a stock room has a given compound on the shelves when there are only a few hundred chemicals listed, no matter what nomenclature is used. If necessary, one can afford to inspect every item individually just to make sure. On the other hand, when the stock room has, let us say, 10,000 chemicals on its shelves it is no longer feasible to search through every item except in case of dire necessity. Unless the nomenclature used in such a stock room is well understood by everyone concerned, considerable time may be spent in searching for the compound, and even then the patron may become discouraged prematurely.

As a general rule, the number of people using a stock room increases roughly with the number of items available. So the possibilities of wasting time because of nomenclature problems increase markedly with the number of compounds on the shelves.

This is true even when the stock room and the users are within the same department and there is an opportunity for the users to become familiar with the stock room setup. The problem is, of course, aggravated when people make requests from other departments by phone. In order to give satisfactory service from a stock room of this type, it is necessary to do at least one of three things: use a single standard name for each compound, have a very large cross reference file, or have someone available who is familiar enough with nomenclature to translate the names back and forth on demand.

Consider a laboratory which is carrying out an evaluation program. If the number of chemicals being tested is in the order of 50 to 500 over a period of 5 years and they are all prepared within the evaluation group, then the chances are that nomenclature presents no problem at all. When the evaluation program involves 5000 or more chemicals and these are being supplied by 20 or more departments, nomenclature may be the cause of some rather expensive situations. For example, there is the problem of samples of the same material submitted by more than one source under different names. Very careful cross-checking is required to make sure the work has not been done before or the work may be done over again. In either case, productive effort is wasted because of nomenclature.

Think of the problem of reporting the results back to the originating laboratories! If the laboratory doing the testing is using a different name from the one used by the originating laboratory, then both names must be associated with the results or more time will be wasted in the process of interpreting the results. This is particularly true if the report is re-examined a year or two after the date of issue. This copying and recopying of multiple names is simple in itself, but in the aggregate it is not only an unnecessary expense but an aggravation to all concerned. When it comes to verbal reporting and the discussion of the results obtained with representatives from other groups within the organization, an appreciable amount of time can be spent in simply explaining the identity of the compounds, time which might be spent to a better advantage on other things.

These two cases are typical of the many ways in which nomenclature problems can enter into the internal communications of an industrial organization.

There is the problem of communicating with people in other organizations. In this case, the problems are similar to those described except that they are more numerous and more difficult to handle efficiently. There are more organizations to send samples to and often each organization has its own peculiar nomenclature practices.

Standardized Nomenclature

Many man-hours could be saved if a single name were used for each compound by everyone concerned. The economics of the situation is forcing the larger organizations to a more standardized nomenclature. By what appears to be common consent, or perhaps common sense, the names used by *Chemical Abstracts* for chemicals are gradually being accepted as the standard names by several of the major chemical organizations.

The *Chemical Abstracts* names have been used in this manner for over 5 years in The Dow Chemical Co. When the idea was first proposed, there were objections at every hand, and many predicted that it could not be made to work. It has been and is working, and everyone has been pleasantly surprised at how much time has been saved by this procedure. The question which is heard now from the research group and in particular from the recent graduates is "Why can't we replace the common names with systematic names?"

There is one phase of nomenclature which is unique with industry and that is the spending of money with the hope of making names better known to the public through advertising. When one is spending $1000 or more on a name, one tends to consider that name much more carefully than when the name is simply to appear in a magazine article which is published free of charge.

Increasing Volume of Scientific Literature

There is still another aspect of nomenclature which is of importance to industry and that is the names and terms as they appear in printed literature. There is no point in reciting here the inadequacies of the nomenclature of our current literature. Untold man-hours are being wasted because of this. One wonders how long it will be before the situation is improved by common effort.

To make matters worse, more material is published today than ever before, and the rate of publication is increasing day by day as well. This rate of increase in the literature is by no means insignificant as can be readily seen from the following facts:

Crane (*1*) stated recently:

The consensus of opinion seems to be that there are at least 600,000 synthetically prepared organic compounds and 30,000 known natural organic compounds. In addition it is estimated that there are apparently 30,000 known inorganic compounds and about 1500 known mineral species. This makes a total of 661,500. As an estimated total a rounder number probably should be used. *Chemical Abstracts* now announces at least 30,000 new compounds each year.

If this present trend is continued, the number of known compounds will be doubled in about 22 years.

Freemont Rider (*2*) states that "university libraries are doubling in size every 16 years." Some of our scientific libraries are growing even more rapidly; they are doubling every 10 years.

In order to get a more concrete idea as to the real significance of this rate, let us assume that our scientific literature has been doubling in size every 10 years since the founding of the AMERICAN CHEMICAL SOCIETY. If the amount of scientific literature in existence in 1876 is represented by the area of a circle 0.4 inch in diameter, then the amount now available would be represented by the area of a circle approximately 4.5 inches in diameter. The amount which will probably be acquired in the next 10 years would, of course, be represented by the area of another circle 4.5 inches in diameter. In other words, the chances are as much information will be accumulated from 1951 to 1961 as was accumulated over the entire period from 1876 to 1951. If the information available in 1876 were doubled it would not be so serious. The volume is now so great, however, that it is becoming increasingly more difficult to take the doubling in stride.

The bill which we are currently paying for our general laxity in nomenclature is already a sizable one. Of course, this amount cannot be calculated accurately, but just to get an idea of the magnitude, let us assume that each member of the AMERICAN CHEMICAL SOCIETY spends only 10 minutes per year on unnecessary nomenclature puzzles. In the aggregate, this would amount to approximately 11,000 man-hours or the equivalent of 5 years of a single individual's normal working time.

Consider the amount of scientific literature which will probably be available 75 years hence. If we continue to use circles and to assume that our libraries will continue to double every 10 years, then the amount of information which will be available in 2026 would be represented by a circle approximately 48 inches in diameter.

The boundary between the known and that which lies beyond is always increasing in size. This means that there will always be an increasingly larger borderline fringe between the known and unknown where there will be more perplexing nomenclature problems than ever, regardless of how well those problems in the better known areas are solved.

Nomenclature plays an important role in the larger industrial organizations and in our literature today. Scientific information is accumulating at an ever increasing pace. Industry is already learning to use nomenclature as a tool and this is being dictated by simple economics. It is only a matter of time before these same economics will force scientists to use a standard nomenclature wherever documentation is concerned.

Literature Cited

(1) Crane, E. J., *Little Chem. Abstracts*, No. 60 (Christmas, 1950).
(2) Rider, Freemont, "The Scholar and the Future of the Research Library," New York, Hadham Press, 1944.

RECEIVED August 21, 1951.

Development of Chemical Symbols and Their Relation to Nomenclature

G. MALCOLM DYSON
Loughborough, England

> The evolution of symbols indicative of elements and compounds may be divided into two distinct periods, the alchemical, when the purpose of the symbols was obscurantism, and the chemical period, when they were employed as a convenience, for clarity and brevity. A short description of some alchemical symbols which have a bearing on later chemical development is given. The chemical symbols of St. F. Geoffroy (1718) and of John Dalton (1808) are discussed, together with the original form of the present-day system of symbolizing elements, as first set out by J. J. Berzelius (1811). The system of Berzelius, as modified by later workers, was found to suffice for simple inorganic compounds, but the growth of organic chemistry led to a demand for structural representation (as opposed to molecular formulation). This was first used in 1858 by A. S. Couper, who laid the foundation for what Butlerow (1860) termed the "structural" and Erlenmeyer the "constitutional" formulas of organic compounds. This pictorial definition of structure had its effect on the naming of organic compounds, and enabled the establishment of a semisystematic nomenclature. The structural formula is internationally and universally understood, but has defects which are not entirely overcome by current nomenclatures; this has led to attempts to overcome such difficulties by establishing methods of codification, and several independent systems have been suggested.

Alchemists used signs or symbols to mystify and keep hidden their secret art. Modern scientists use symbols as a convenience, as abbreviations saving time and thought, but so far removed from attempts at secrecy that they will admit of no system unless it be clear, and demonstrates both conciseness and precision. When this change took place is not clear, for it was a gradual one. As J. W. Mellor said: "Language generally lags in the wake of progress" (*13*). We find the eminent physician John Quincy (*15*) in 1730 publishing a list of symbols used by physicians in writing their prescriptions. While these abbreviated forms, shown in Table I, are undoubtedly a convenience, some remnant of the older purpose may yet have remained—namely, to prevent the patient from knowing the nature of the medicine. To quote from Quincy himself on this point:

   Abbreviations in Medicine are certain Marks or half Words used by Physicians for the sake of Dispatch and Conveniency in their Prescriptions, tho' some are pleas'd to give another Interpretation to the thing, as if it was design'd to conceal their Art from such as knew less *Latin* than themselves, or their ignorance from such as know more: but this

Table I. John Quincy's "Characters" (1730)

♁	Antimony	♃	Nitre	Cong.	Gallon
AF	Aqua Fortis	※	Sal Ammoniac	gr.	Grain
AR	Aqua Regia	☾	Silver	ƒs.	Half any thing
MB	Balneum Mariæ	♏	Spirit of Wine	M	Handful
♀	Calx Viva	△	Sulphur	℥	Ounce
⊖	Common Salt	S.V.R.	S.V.R. (Spir. vin. rect.)	℔	Pound or a Pint
♀	Copper	⊔	Tartar	P	A Pugil (pinch)
☉	Gold	♃	Tin	Ͻ	Scruple
CC	Harts-Horn	⊕	Vitriol	Cochl.	Spoonful
CCC	Harts-Horn Calcined	qs	A Sufficient Quantity	☹	Caput Mortuum
♂	Iron	SA.	According to Art	↷	Distill
♄	Lead	ʒ	Drachm	☿	Precipitate
☿	Mercury	PÆ	Equal Quantities	☿	Sublimate

kind of Short-hand is very convenient in urgent Cases, or where a Patient's life might be lost whilst a Man could write half a sheet in the long way.

Most of Quincy's signs are zodiacal in origin, or just plain abbreviations of the Latin form, and some are still in use.

The death of the phlogiston theory and the gradual emergence of the concept of constant composition forced upon chemists the realization of the pressing need for a system of precise nomenclature which would:

Indicate the compound.
Define it.
Recall its constituent parts.
Classify it according to its composition.
Indicate the relative proportion of its constituents.

These were held to be the attributes of the system of Greyton de Morveau, Lavoisier, Berthollet, and Fourcroy (9) which, indeed, introduced inorganic nomenclature in approximately the form in which we now use it. They introduced the terminations -ic and -ous to indicate the valence states—although they themselves would not have expressed it as "valence"—of such elements as copper and iron, and "-ate" and "-ite" for salts of different acids of the same series. Berzelius (1) in 1811 consolidated this scheme and introduced the termination "-ide" for combinations of 2 elements.

Table II. John Dalton's "Characters"

⊙	Barîtes	⊙	Mercury	⊕	Sulfur
●	Carbon	◐	Nitrogen	Ⓩ	Zinc
©	Copper	○	Oxygen		
Ⓖ	Gold	⊕	Phosphorus	⊙⊙	Water
⊙	Hydrogen	Ⓟ	Platinum	○●○	Carbon dioxide
Ⓘ	Iron	⦀	Potash		Sulfuric acid
Ⓛ	Lead	Ⓢ	Silver		
⊙	Lime	⦵	Soda	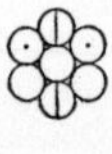	Ammonium nitrate
⊛	Magnesia	⊙	Strontites		

These systems could scarcely have prospered without the atomic theory of Dalton, and there is little doubt that Dalton's promulgation of the atomic theory led to a new epoch in both symbolization and nomenclature (*3*). Dalton himself had a series of symbols, shown in Table II, clearly based on the older types and quite arbitrary in their selection.

The shape of things to come is seen both in the use of the following symbols for the Aristotelian elements used by the 13th century alchemists:

△ Fire
△ Air
▽ Water
▽ Earth

and in Lully's use of substituted letters (*12*).

A Gold
B Mercury
C Saltpeter

For nearly half a millennium, confusion arose from varying schemes of symbolization. Each author adopted his own whims, although a broad overlapping similarity was observed with many of the systems. Geoffroy (*8*) used the current systems with his own additions to portray chemical reactions.

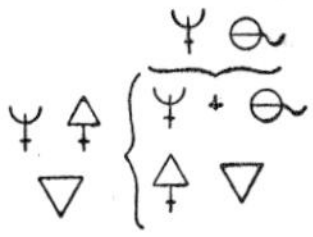

$$CaS + H_2SO_4 (H_2O) \longrightarrow CaSO_4 + (H_2O)$$

This indicates that calcium sulfide solution ♆ ♃ ▽ (the ▽ is the water sign persisting from time immemorial) with sulfuric acid ☉ gives sulfur △, calcium sulfate ♆ ☉, and water ▽.

It needed a flash of genius by Berzelius in 1811 (*1*) to realize that the simplest and most mnemonic symbol for an element was the initial letter of its name with a second lower case letter for characterization where necessary. Common elementary symbols are but 140 years old, and even Berzelius did not use them in the modern way. He drew a line through the symbol to indicate two atoms—e.g., C̶u̶-O or H̶-O for our Cu_2O and H_2O and further abbreviated by representing oxides and sulfides by dots or primes superimposed upon the elementary symbol:

H_2O	Ḧ̶
Fe_2O_3	F̈e̶
$CaOCO_2(CaCO_3)$	CaC̈
$CuSO_4 \cdot 5H_2O$	Ċu′,5Ḧ
FeS	Fe′
$FeSO_4$	F̈e′

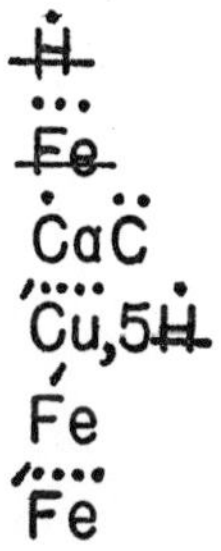

This system, which sacrificed clarity to brevity, was soon dropped by chemists in favor of the subscript numeral style now universal, but it persisted in mineralogy for many years.

There is no doubt that, as these formulas became more widely and generally employed, they had a profound effect on nomenclature. The immediate effect was to relegate the trivial names for inorganic substances and to substitute logical and systematic names. No longer did chemists talk of green, blue, and white vitriols but of ferrous, cupric, and zinc sulfates. What Dumas (4) called *la langue des cuisinières*—butter of antimony, butter of zinc, and butter of arsenic or the oils of vitriol, tartar, and olive—was replaced by more modern terminology: antimony trichloride, zinc chloride, arsenic trichloride, sulfuric acid, etc. In other words, where the common denominator of groups of analogously named substances (butters, oils, vitriols, spirits, ethers, aquæ, tartars, and salinæ) had been a superficial external physical resemblance, it now became a logical and universal reference to chemical composition. This rational nomenclature had a refining and clarifying influence on chemical thought, enabling an almost intuitive correlation between name, formula, and composition. As A. Laurent (11) remarked in 1854:

> For a language to be perfect, it is not sufficient that each substance, each idea, each modification of form, time, place, etc., should be represented by one word, or by one invariable symbol; it is necessary in addition, both to aid the memory and to facilitate the operations of the mind, that analogous words should designate analogous substances, analogous ideas, and modifications of ideas. It is thus that the words of our language represent to us by similar terminations or augments similar modifications of ideas represented as when we say: *je vois, j'aperçois, je reçois; nous voyons, nous apercevons, nous recevons.* In like manner do chemists make use of the expressions sulphate, nitrate, chloride, etc.

The one-one correspondence between symbols (formulas) and nomenclature is very highly developed in inorganic chemistry, and played a great part in the development of the subject. It is only when the more complex ionic structures must be named that it fails and group or radical nomenclature must be substituted. This is an important development, even though it seems simple in retrospect. In establishing an equivalence between:

CaS	Calcium sulfide
$FeSO_4$	Ferrous sulfate
$CoCl_3$	Cobalt trichloride
NaH_2PO_4	Sodium dihydrogen phosphate

radical nomenclature has been tacitly assumed to be permissible and the foundations of ionic nomenclature unconsciously laid. Coming to $Co(NH_3)_6Cl_3$, it is clear that such a substance differs from either $CoCl_3.6NH_4Cl$, or a hypothetical double salt $Co(NH_4)_6Cl_9$ by some factor that is greater than the mere loss of 6 molecules of hydrogen chloride—but the principle of one-one correspondence between formula and name is restored when the compound is written ionically as $[Co(NH_3)_6]Cl_3$ and named hexamminecobalt trichloride.

Chemists who remember the "preionic" nomenclature with its "purpureo, praseo, roseo, luteo, fusco, and xantho" nomenclature will agree that not only is the newer method simpler to apply, clearer in its implications, and less in its demands on sheer memory, but it suggests analogies and "facilitates the operations of the mind" in precisely the manner

Table III. Fownes' List of Radicals

Amyl (symbol Ayl)	$C_{10}H_{11}$	Benzoyl (symbol Bz)	$C_{14}H_5O_2$
Amyl-ether	$C_{10}H_{11}O$	Hydride of benzoyl;	$C_{14}H_5O_2H$
Hydride of amyl	$C_{10}H_{11}H$	bitter almond oil	
Potato-oil	$C_{10}H_{11}O, HO$	Hydrated oxide of benzoyl;	$C_{14}H_5O_2O, HO$
Chloride of amyl	$C_{10}H_{11}Cl$	benzoic acid	
Bromide of amyl	$C_{10}H_{11}Br$	Chloride of benzoyl	$C_{14}H_5O_2Cl$
Iodide of amyl	$C_{10}H_{11}I$	Bromide of benzoyl	$C_{14}H_5O_2Br$
Zinc-amyl	$C_{10}H_{11}Zn$	Iodide of benzoyl	$C_{14}H_5O_2I$
Acetate of amyl	$C_{10}H_{11}O, C_4H_3O_3$	Sulphide of benzoyl	$C_{14}H_5O_2S$
Sulphamylic acid	$C_{10}H_{11}O, 2SO_3, HO$		
Amylene	$C_{10}H_{10}$		
Valerianic acid	$C_{10}H_9O_3, HO$		

Table IV. Couper's Structural Formulas

$$\begin{array}{l}\text{C}\ldots\text{O}\ldots\text{OH}\\ \text{C}\ldots\text{H}_2\\ \text{C}\ldots\text{H}_2\\ \text{C}\ldots\text{H}_3\\ \text{Propyle alcohol}\end{array} \qquad \text{C}\begin{cases}\text{C}\ldots\text{H}_2\\ \text{C}\ldots\text{H}\end{cases} \qquad \text{C}\begin{cases}\text{C}\ldots\text{H}\\ \text{C}\ldots\text{H}\end{cases}$$

$$\text{C}\begin{cases}\text{C}\ldots\text{H}\\ \text{C}\ldots\text{O}\;\text{OH}\end{cases} \qquad \text{C}\begin{cases}\text{C}\ldots\text{H}\\ \text{C}\ldots\text{O}\ldots\text{O}\end{cases}$$

$$\text{C}\begin{cases}\text{O}_2\\ \text{O}\ldots\text{OH}\end{cases} \qquad \text{C}\begin{cases}\text{O}_2\\ \text{O}\ldots\text{O}\end{cases}\Big\}\text{PhCl}^5$$

Salicylic acid   Terchlorophosphate of salicyl

envisaged by Laurent. The Berzelius dot and prime method was shorter than the subscript numeral system universally adopted; xanthocobalt salts is shorter than nitritopentamminecobalt salts; instances of this kind can be multiplied indefinitely to illustrate that clarity must be first, brevity second.

The principle of one-one correspondence between empirical formula and name can only be useful with the simplest organic substances. Carbon tetrachloride, CCl_4, or carbon tetrabromide, CBr_4, are examples, but if the principles of the inorganic nomenclature-symbol relation are carried further the result is:

$CHCl_3$ Carbon hydrogen trichloride
C_2H_6 Dicarbon hexahydride

There was no way out of this dilemma, save trivial names, until knowledge of structural formulas was sufficiently advanced for chemists to use them as the basis of nomenclature. Thus, without structural symbolization, systematic organic nomenclature could not arise. The radical theory was the first step in the direction of systematization, but even in 1852 the detection of a common factor as shown by Fownes (see Table III) (7) was unable to do much toward solving the problem.

The first communication to employ structural formulas was one by Couper in 1858 (2). His tentative formulas with dotted lines (Table IV) are the forerunners of our modern structural formulas, while the conception of the ring established by Kekulé (10) was needed to place the keystone in the whole system. As soon as structural formulas became established, systematic nomenclature could follow, and when it was possible to write:

$(CH_3)_2CHC(Et)_2OH$ in place of $C_8H_{17}OH$

then the form 3-ethyl-2-methylpentan-3-ol could follow. When [benzene ring with HO, NH_2, CH_3 substituents] was available, 1-methyl-4-hydroxy-2-aminobenzene could automatically be written. It would seem, therefore, that once an invariant structural formula had been established, then a unique systematic name would follow. Unfortunately, this is not true for many reasons, chiefly:

Persistence of trivial and semitrivial (or semisystematic) names—e.g., nitrotoluidine, aminocresol, etc.—which, though convenient, are not systematic and may be confusing, as, for example, the nitroanisidines, where *o*-nitro-*p*-anisidine is an ambiguous form.

The absence of agreed rules for name forming (partly remedied by the various conferences of the Commission on Organic Nomenclature of the International Union of Pure and Applied Chemistry).

As a principal factor, the loss of a strict one-one correspondence between structure and nomenclature so marked and so valuable in inorganic nomenclature.

Nomenclature chaos in organic chemistry will not be ordered until a one-one correspondence between structure and name is restored. It was with this object specifically in view that a notational system (5) was devised which will generate a unique systematic name in every case. This does not, of course, mean (as some have supposed) that the sole and exclusive use of systematic names for organic substances is advocated. It would be pedantic to talk of hydroxybenzene or 4-hydroxy-1-methylbenzene when phenol or *p*-cresol is meant. But there must be an agreed, unique systematic or fiducial name for

Table V. Notation System for Organic Compounds
(Dyson, 3)

Formula	Notation	Systematic Name	Common Name
CH_3OH	C.Q	Methane-ol	Methanol
CH_3CH_2OH	C_2.Q	Ethane-ol	Ethanol
$CH_3CH_2CH_2OH$	C_3.Q	Propane-ol	Propanol
$CH_3CHOHCH_3$	C_3.Q,2	Propane-2-ol	2-Propanol
 (phenol structure)	B6.Q	Hexaphene-ol	Phenol
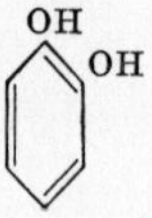 (catechol structure)	B6.Q,1,2	Hexaphene-1,2-diol	Catechol
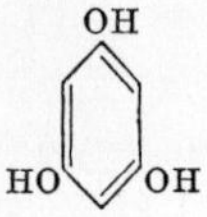 (phloroglucinol structure)	B6.Q,1,3,5	Hexaphene-1,3,5-triol	Phloroglucinol
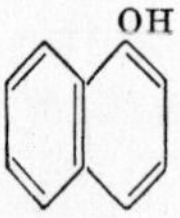 (α-naphthol structure)	$B6_2$.Q,3	Binihexaphene-3-ol	α-Naphthol
$(CH_3)_2C(CH_2)_2CH(Et)CH_2-CH(CH_2CH_2CH_3)CH_2CH_2CH_3$	$C_{10}.C_3,4.C_2,6.C,9,9$	Decane,4-propyl,6-ethyl,-9,9-dimethyl	....
 (phthalazine structure)	$B6_2ZN,4,5$	Binihexaphene-4,5-diaza	Phthalazine
$CH_3CH_2CH(Me)CH_2CHO$	C_5.C,3.EQa	Pentane, 3-methyl-1-al	....
$CH_2ClC(CH_2Cl)_3$	$C_3.C,2,2.Cl,1,3,4,5$	Propane-2,2-dimethyl-1,3,-4,5-tetrachloro	1,3,2',2'-Tetrachloro-2,2-dimethylpropane

every structure in the background, to be called on when needed (as it is so often, in the naming of new classes of compounds). Foster (6) wrote:

In forming the nomenclature of any science, two distinct requirements must be kept in view ... a convenient general language ... for everyday use; and what may be termed the legal language of science ... where terms are strictly defined to have an exact and generally recognized value.

Patterson (14) underlines this when, in referring to notations, he states:

None of these, however, can take the place of a system of good names which can be used in writing and speaking and in subject indexes. There is, however, no reason why a system of notation, such as that of Dyson (5) should not be used to engender a system of fiducial nomenclature.

It has, therefore, been the aim in developing this system of notation to keep well in mind this prime necessity, and to use only symbolization which can generate unequivocal systematic names. In Table V a series of compounds is shown in compared nomenclature. These names (with the possible agreed elision of e- wherever it occurs) can serve as systematic names. The simplicity of the system and its ability to generate names are illustrated in Table V.

This interconvertibility of structure and name is invaluable to the organic chemist. It follows his normal line of thought, produces names that are largely familiar in connotation, and requires little effort to use. There will be a long list of exceptions to the general or colloquial use of such names, but so there will be with any formal system, for Foster's dictum (6) on the two languages necessary for the science is still true.

Two facts stand out from all experience in notation and nomenclature: the urgent necessity for a rigid systematic nomenclature for organic compounds and the essentiality of a one-one correspondence between structure, notation, and nomenclature. Perhaps no system of notation or nomenclature will be other than a dead letter unless it follows the natural line of thought of the user, and does not depart too much from his common habits of expression. That great scientist, Lord Rayleigh, said:

Science is nothing without generalizations. The suggestion of a new idea, or the detection of a new law, supersedes much that has previously been a burden to the memory, and, by introducing order and coherence, facilitates the retention of the remainder in an available form.

Such a service will, it is hoped, be performed for organic chemistry by the future developments of notation and nomenclature in concert.

Literature Cited

(1) Berzelius, J. J., *Am. Phil.*, **3**, 51, 363 (1814); "Lehrbuch der Chemie," Dresden, 1827.
(2) Couper, A. S., *Compt. rend.*, **46**, 1157 (1858).
(3) Dalton, J., "New System of Chemical Philosophy," 2 vols., London, 1807–10.
(4) Dumas, J. B. A., from reference (*13*), p. 119.
(5) Dyson, G. M., "New Notation and Enumeration System for Organic Compounds," 2nd ed., New York, Longmans, Green and Co., 1949.
(6) Foster, G. C., *Phil. Mag.*, **29**, 266 (April 1865).
(7) Fownes, G., "Manual of Elementary Chemistry," 4th ed., pp. 463, 473, London, 1852.
(8) Geoffroy, St. F., "Table des diffèrents rapports observés en chimie entre diffèrentes substances," 1718.
(9) Greyton de Morveau, L. B., Lavoisier, A. L., Berthollet, C. L., and Fourcroy, A. F. de, "Méthode de nomenclature chimique," Paris, Cuchet, 1787 (London, 1799).
(10) Kekulé, A., *Bull. soc. chim.*, **3**, 98 (1865); *Ann*, **137**, 129 (1866).
(11) Laurent, A., "Méthode de Chimie," Paris, 1854.
(12) Lully, Raymond, "Testamentum, duobus libris universam artem chimicam complectens," Cologne, 1568.
(13) Mellor, J. W., "Comprehensive Treatise on Inorganic and Theoretical Chemistry," Vol. 1, p. 119, London, Longmans, Green and Co., 1922.
(14) Patterson, A. M., *Chem. Eng. News*, **29**, 3036 (1951).
(15) Quincy, J., "New Medicinal Dictionary," London, J. Osborn and T. Longman, at the Ship in Pater-noster Row, 1730.

RECEIVED November 1951.

The Role of Terminology in Indexing, Classifying, and Coding

JAMES W. PERRY[1]
Massachusetts Institute of Technology, Cambridge, Mass.

> The role of terminology in establishing alphabetical indexes; in defining the subjects or ideas used to group information when fixed classification schemes are used; and in defining concepts which are indicated physically on punched cards is discussed. Advantages and disadvantages of alphabetical indexes, classification systems, and punched cards are enumerated.

Terminology and nomenclature serve science and technology in a variety of ways, of which the most obvious are the communication and recording of information. Equally important is the use of scientific and technical terms as building blocks for constructing indexes, classification systems, and codes for mechanical searching.

Different meanings are sometimes attached to the words: indexing, classifying, and coding.

In this paper indexing is defined as a system which provides leads to needed information by the three steps—selection of appropriate terminology (most of which is usually found in the material being indexed); modification of the selected terms by suitable words and phrases; and alphabetical arrangement of the selected terms and their modifications.

Classification is defined here as an arrangement whereby various similar items are grouped together. For example, printed copies of patents, related by subject, may be grouped in pigeonholes. An alternate method of classification is that of arranging books in a fixed order, on the basis of a predetermined system for indicating the subject matter treated in the books. In actual index building, a certain amount of classification obtained by wording and arrangement of entries has been found helpful to the user.

The word coding has been used on occasion, and is used here, to mean a method of analyzing information in terms of various aspects. In order that such analysis may be used to fullest effectiveness, it is essential that there be some physical means for separately indicating various characteristics and aspects of the information being analyzed, so that search can be directed to any one aspect or any combination thereof. At the present time, one of the most convenient means for accomplishing this is punched cards (6).

The purposes which indexing, classifying, and coding serve provide a convenient starting point for considering the role of terminology. Perhaps the simplest purpose is the recall of recorded facts. For example, an index may be used to look up the melting point of some single compound. A classification system may be used to locate patents describing the synthesis of azo dyestuffs. A punched-card system may be used to separate out those U. S. patents concerned with the uses of kerosene-derived alkylaryl sulfonates.

Over and above simple recall—and yet closely related thereto—the different systems for analyzing information may be used as a basis for effecting correlations. Information

---

[1] Present address, Battelle Memorial Institute, Columbus, Ohio.

may be selected to establish such relations as cause and effect or—by using statistical analysis—to establish relationships between the probability of one event and the occurrence of another. Indexing, classifying, and coding may also serve as a basis for building up more complex networks of relationship. An example is searching out a series of reactions for synthesizing a complex molecule. Since the establishment of generalizations—in the form of hypotheses, theories, and natural laws—is one of the most powerful methods for gaining and applying scientific insight, the role played by indexing, classifying, and coding in the advance of science is an important one.

But regardless of whether information is being sought out in order to plan an experimental program or to provide factual background for a new theory, the basic operation in using scientific and technical literature is selection of pertinent items. The selecting operation, if truly efficient, will avoid time being wasted in considering the nonpertinent. Effectiveness in accomplishing rejection of the unwanted is the touchstone for testing the efficiency of the different systems for searching out needed information.

The use which is made of terminology in indexing, classifying, and coding is determined, to a large degree, by practical considerations which arise from the different mechanical devices used as adjuncts to searching and correlating operations (10).

Indexing

The basic device used in indexes for achieving rejection of the unwanted is the alphabetical arrangement of index entries. An index performs the rejecting function in an ideal fashion when the needed information is located under a single heading. This occurs, for example, in using the telephone directory when there is no doubt concerning the spelling and alphabetizing of a name. An index also functions in an ideal fashion when information about a single compound is required, and there is no uncertainty of nomenclature and alphabetizing. In such an instance the alphabetical arrangement of index entries is a swift and reliable guide to the needed information. Generally, alphabetized indexes are at their best when the required information is under one heading that can be recalled easily.

Rejection of the unwanted is no longer ideal when uncertainty arises as to the appropriate heading for locating needed information. A simple example of such uncertainty is provided by alternate spelling, such as Thomson instead of Thompson in the telephone directory, or by synonyms, when only one is used, as illustrated by the cross reference, "Quicksilver. See Mercury."

Individual alternative terms cause relatively little difficulty in comparison with the difficulties arising when the information needed is not restricted to a single person or compound. Often, in using the chemical literature, the characteristics of several related compounds or of a class of compounds are required. In this situation, an alphabetized index may compel the examination of a number of different headings. Searching around for all of the compounds making up a class can, on occasion, become quite burdensome and may prove to be so laborious that it constitutes a task which is economically impossible to undertake.

Those who construct indexes are well aware of the desire on the part of index users to search for a more-or-less extensive array of closely related substances (2). It is for this reason that cross references and a certain amount of classification are built into the subject indexes of *Chemical Abstracts*. Thus, under the heading "Glucose" are included alpha and beta forms, the solutions of glucose, analogs of glucose, containers for glucose, etc. Under "Iron"—to give another example—is found information pertaining not only to the metal but also to its ions, its compounds, iron substitutes and the like, as well as books and journals on iron, and studies on the word itself. A device frequently used to incorporate a degree of classification in subject indexes is the inversion of words and terms. Thus, chloroacetic acid is made to appear in the index near "Acetic acid" by alphabetizing in the form "Acetic acid, chloro." Similarly, electric insulators are alphabetized in the indexes of *Chemical Abstracts* as "Insulators, electric," and thus are grouped with a variety of other insulators. Another method for drawing together related items in an index is to enter closely related or easily interchangeable things under one heading. For example, pH and hydrogen-ion concentration are grouped by *Chemical Abstracts* under "Hydrogen-ion concentration." Terms defining broader classes may be used as a cross reference to effect classification. *Chemical Abstracts* uses the cross reference, "Magazines. See Literature." Another very important type of classification is obtained by the grouping of related ideas under index headings, for example, "Aluminum, reaction with HCl; reaction with NaOH; reaction with H_2SO_4."

Practical considerations prevent indexers from achieving more than a limited amount of classification. It might, in theory, be possible to list under olefins all unsaturated hydrocarbons. If this were done for every conceivable generic term, such as olefins, an index would be inflated to such size that its mere bulk might prove an impediment to its use, if the cost of compiling and printing were not sufficient deterrents to such exhaustiveness. As a consequence, under the broad generic terms, such as "Olefin," "Carbohydrates," and "Sugar," *Chemical Abstracts* enters only those papers which pertain to the general subject and omits reference of those papers which pertain to individual members of such classes of compounds. This means, of course, that the generic term "Olefin" is not effectively exploited as a means for providing leads to those papers concerned with individual hydrocarbons of that general class. Inability of subject indexes to exploit generic terminology fully is one of their major limitations.

Another important limitation is often encountered when making searches that can be defined only by using a combination of terms, as for example "Glucose, as binding agent for pills." In constructing indexes, the user's interest in being able to search for combinations of concepts is, of course, not overlooked. The majority of the index entries of *Chemical Abstracts* consist of at least two concepts tied together to form a more or less complex phrase. Here, however, practical limits are encountered. The entries would become excessively lengthy and difficult to scan if each index entry cited all those concepts that were interrelated in the original paper or its abstract. A selection must be made from the numerous possible combinations of concepts that might be set up as index entries. Such selection is made by the indexer in line with his expectations as to the user's requirements. In making these selections, it is not possible to take into account unexpected future trends in scientific discovery and technical development. Such new combinations not anticipated by the indexer may and, in fact, do present the future user of the index with difficult—or, at least, time-consuming—problems. The alphabetized index cannot furnish a complete answer to the problem of searches definable only in terms of combinations of concepts. The situation becomes particularly difficult when one or more general terms are required to define the scope of the search, e.g., compositions containing surface active agents and dyestuff for synthetic fibers.

From the user's point of view, searching out an individual bit of information in an index will be easy, provided the word or phrase used by the indexer as a guide suggests itself to the searcher, or can be picked up through a cross reference. On the other hand, searching for a range of information becomes more difficult as the number of index entries involved increases, or as the difficulty of imagining the terminology used to index various members of the class increases. In general, an index is at its best, in comparison with other methods for locating information, whenever the scope of the search can be defined precisely in terms of a single index entry. This means that indexes are particularly well suited for searches relating to the various individuals, be they telephone subscribers, chemical compounds, or other specifically named entities. The limitations of indexes arise from inability to provide for all possible combinations of concepts and lack of full exploitation of generic concepts. In particular, searches defined by combinations of concepts are difficult to accomplish with conventional indexes.

Classification

Classification, as exemplified by the system employed by the United States Patent Office, groups together in fixed compartments items having certain common features. In the Patent Office classification, the subject matter of the patent is analyzed by the classifier in terms of the features on which the classification system is built. Terminology ranging from the highly generic, e.g., carbon compounds, toys, medicines, poisons, and cosmetics, to the highly specific, e.g., thiophene, marionette, nicotine, thyroid, is used to specify the different features which are the basis for deciding into which compartment a given patent is to be placed (*3, 4*). This becomes more clearly evident when excerpts from a classification scheme are considered (*9*).

    260 Chemistry. Carbon Compounds
          239 Heterocyclic carbon compounds
          313   Pyrroles
          319   Indoles
          321   Indigoid
          322   Thiophene nucleus
     46 Amusement Devices, Toys
          115 Figure toys
          119   Movable
          126   Marionette
    167 Medicines, Poisons, and Cosmetics
          13 Poisons
          22   Organic
          33   Heterocyclic compounds
          34   Nicotine

    The Patent Office classification system undertakes to accomplish the all-important function of rejecting unwanted items by making it possible to pass over all the groupings—i.e., the classes and subclasses—except the one in which items containing the desired information will be found. Such classification functions ideally when all of the items in a given grouping contain pertinent information while no other items are of interest. In this case, merely selecting the correct grouping (class or subclass) accomplishes the search. In practice, this ideal is rarely realized. Even if it is not necessary to go beyond one grouping some of its items turn out, on inspection, to be of no interest. The designers of classification systems strive to set them up in such a way that only a reasonable amount of time is required for personal inspection of nonpertinent material.

    Classification as practiced by the Patent Office is particularly effective in meeting information requirements that are defined by a generic term, e.g., toys, poisons, heterocyclic compounds. With such searches, the irrelevant is quickly eliminated, and a single grouping may provide the needed material in convenient preassembled form. A search directed to a single item, e.g., some individual poison or compound, may require that a number of items within some one grouping be inspected personally to select the required information from nonpertinent material. Classification is better suited for making relatively broad searches than for pin-pointing some single fact. Since incoming patent applications often contain rather broad claims, it is perhaps understandable that the examiners in the U. S. Patent Office prefer to work with a classification system rather than with an index (*8*).

    The basic limitation inherent in conventional classification is the inflexibility of a rigid system of compartmentalization. Once a large number of patents have been sorted out according to one classification system, rearrangement in accord with some other system requires much effort on the part of some person having expert knowledge of the subject matter.

    Another important limitation is the practical necessity of limiting the number of groupings that are established. In practice this works out so that only a small portion of the totality of logically valid classes and subclasses are set up. Dyestuffs, for example, might be classified on the basis of their chemical constitution, materials which they are used to color, methods of application, and the color imparted. When the classification is affected on the basis of chemical constitution, then a person who may be interested in dyes as chemical substances finds that his purpose is served very well. On the other hand, such classification may provide little assistance to a searcher who is interested in finding all the dyes used to color a certain type of material such as cellulose ester textiles. In establishing a classification system based on fixed compartments, arbitrary decisions must be made as to the basis for classification. If such decisions are in line with the user's requirements, his needs will be well served, particularly for making broad generic searches; if, on the other hand, a search must be conducted that is completely out of line with the basis of classification, then the searcher may be aided relatively little by detailed breakdown into subclasses. Furthermore, when a single item, e.g., a single dyestuff, is the subject of the search, it may or may not be relatively easy to locate by working through a classification system. If the classification system has been drawn up so the desired item will be located in one specific subclass, then, usually, relatively little time is required to examine that subclass. Sometimes the classification system does not permit certain predictions that a needed piece of information will be found in one specific pigeonhole or subclass. It may often be necessary to search through several of them.

The art of conventional classification is to harmonize, as well as possible, the unavoidable arbitrary decisions as to basis of classification with the needs of the persons using the system. The more variable these needs, the greater the difficulties encountered.

Conventional classification uses terminology to establish and define classes, subclasses, and sub-subclasses. Distinction between a broad class and narrower subclass may be made by using progressively less generic terms as in the example cited from the U. S. Patent Office classification (9).

```
313    Pyrroles
319       Indoles
321          Indigoid
```

Alternately two or more generic terms may be used jointly.

```
13    Poisons
22       Organic
33          Heterocyclic compounds
```

Here subclass 22 is defined as those poisons that are organic substances; the scope of the subclass is the intersect of the two broader categories. Similarly two additional generic terms, "heterocyclic" and "compound," are used to relegate certain organic poisons to sub-subclass 33. With these terms an entirely different grouping might have been set up, for example:

```
13    Organic compounds
22       Heterocyclic
33          Poisons
```

Or alternately:

```
13    Organic compounds
22       Poisons
33          Heterocyclic
```

This mere rearrangement of terminology would result in an entirely different set of items being embraced within each of the various groupings.

Practical necessity forbids establishing all these alternate groupings, as the number of fixed compartments cannot be increased without making the system more costly to set up and maintain. Each alternate combination of terms employs them in a different way, that sooner or later can be expected to be useful. Thus the exploitation of terminology by conventional classification is restricted by practical considerations which have their origin in the use of fixed compartments.

A similar situation exists in the closely related form of classification that arranges items in a monodimensional array, as exemplified by the library classification of books on shelves.

Coding

Alphabetical arrays (as used for indexing) and fixed compartments (as used for classification) are not, of course, the only means for employing terminology to enable desired items of information to be selected. During recent years hand-sorted punched cards have been widely used (*6*). These cards are supplied by the manufacturer with rows of holes punched along the periphery. Clipping out the cardboard between the edge of the card and a hole produces a slot with the result that the card will drop out when sorting. In this way, the clipping of different holes can be used to designate different search criteria. For example, one hole can be clipped for poison, another for organic compound, and another for heterocyclic compound.

A file coded and punched in this fashion can be searched for any single criteria. A search on poisons will select all cards bearing information on such substances, with no differentiation of inorganic from organic or organic from heterocyclic compounds. Similarly all cards bearing information on organic compounds or heterocyclic compounds can be selected, without regard to their coding as being poisons or not. Once all cards on poisons have been isolated, a second sort for heterocylic compounds will separate the heterocyclic poisons. This ability to search to any one characteristic or to their combinations makes it possible to establish, in effect, a specially defined subclass at the time when such a subclass is needed.

Qualitatively, hand-sorted punched cards overcome previous limitations inherent in alphabetized indexes and in classification by compartmentalization. Quantitatively, these cards leave much to be desired. The number of holes on the periphery is small with respect to the total number of concepts needed to code broad ranges of subject matter. It is true that coding of a single concept may be based on any one of the virtually unlimited number of combinations that may be generated from a modest number of individual holes. When this is done, however, the punching and sorting operations become more complicated. With loss in simplicity of operations, something akin to diminishing returns sets in, and this establishes a somewhat elastic though nonetheless effective ceiling on the number of terms that can be encoded independently. Furthermore, as the number of hand-sorted cards in a file increases, it tends to become more awkward to use. As a consequence, the use of hand-sorted punched cards for selecting information has been restricted to relatively narrow fields of specialization.

The problem of conducting searches involving broad ranges of subject matter in amounts generated by present-day research activities cannot be alleviated by halfway measures (*11–13*). Fortunately, electronic devices developed during recent years are able to perform many operations previously regarded as requiring mental effort. Among these operations are comparing and identifying, recording and recalling from memory, adding, subtracting, multiplying, and dividing, and establishing logical sums, products, and differences. Units for performing these operations are the components from which machines designed for specialized purposes have been constructed. Machines can be built to carry through routines no matter how complicated they may be. Modern electronic techniques have revolutionized the art of mathematical computation. The means are at hand for accomplishing equally important advances in searching and correlating other types of information.

Practical considerations require that searching machines be designed so as to eliminate all unneeded complexity. Equal consideration must be given to the preliminary processing of documents so that their subject matter can be searched by machine. This task must be performed by people, at least for the foreseeable future, and consequently it must not exceed human ability. These two problems, machine design and indexing for machine searching, are interlocked. The key to their solution is the systematic exploitation of terminology to indicate those features of subject matter of interest as reference points for defining and conducting searches. Terminology will be needed by the indexer to analyze subject matter appropriately. The scope of searches will also be expressed by using terminology. Its role is, therefore, to link machine operations both to search requirements and to the material undergoing search (*5*).

The indexing operation must result in relating to the subject matter of documents those terms required for searching purposes. Key words and phrases in a given document can be set up as index entries without involving much difficulty. Requiring an indexer to supply a large number of additional headings, particularly of generic nature, would undoubtedly prove impractical. For example, on indexing "ethylene" it would be difficult and tedious to provide such additional terms as "olefin," "hydrocarbon," and "organic compounds," or to indicate, when using "clock" as an index entry, that this term refers to a "device" for "measuring time." Systematization of terminology provides a basis for establishing codes which link specific terms to generic semantic factors (1, 7). In this way the clerical step of encoding—which could be performed by automatic equipment—translates specific terms into a form of representation which makes generic terms available as reference points for defining and conducting searches. Methods for establishing such systematization are under development and will be reported in due course.

Literature Cited

(1) Bernier, C. L., "Relations between Words"; Perry, J. W., "Comments on Relations between Words," privately circulated memoranda, 1948.
(2) Bernier, C. L., and Crane, E. J., *Ind. Eng. Chem.*, **40,** 725–30 (1948). Indexing Abstracts.
(3) Bailey, M. F., "Classification of Patents," 2nd rev., Washington 25, D. C., U. S. Patent Office, 1946.
(4) Bailey, M. F., *J. Patent Office Soc.*, **28,** 463–507, 537–75 (1946). A History of the Classification of Patents.
(5) Bailey, M. F., Lanham, B. E., and Leibowitz, J., in "Abstracts of Papers, 120th Meeting, AMERICAN CHEMICAL SOCIETY, New York," p. 2F. Mechanized Searching in the U. S. Patent Office.
(6) Casey, R. S., and Perry, J. W., "Punched Cards. Their Applications to Science and Industry," New York, Reinhold Publishing Co., 1951; pp. 449–67, Vol. 8 of Kirk and Othmer, "Encyclopedia of Chemical Technology," New York, Interscience Publishers, 1952.
(7) Kent, Allen, Berry, M. M., and Perry, J. W., "Systematization of Terminology," paper presented before the Division of Chemical Literature at the 122nd Meeting of the AMERICAN CHEMICAL SOCIETY, Los Angeles, Calif.
(8) Lanham, B. F., *Ind. Eng. Chem.*, **43,** 2494–6 (1951); *J. Patent Office Soc.*, **34,** 315–23 (1952).
(9) "Manual of Classification," Washington 25, D. C., U. S. Patent Office, 1947. (Continuing revisions.)
(10) Perry, J. W., *American Documentation*, **1,** 133–9 (1950). Information Analysis for Machine Searching.
(11) Pietsch, E., *Chimia*, **7,** 49–57 (1953). Aus der Arbeit am Gmelin-Handbuch der anorganischen Chemie.
(12) Pietsch, E., "Future Possibilities of Applying Mechanized Methods to Scientific and Technical Literature," in Casey and Perry, "Punched Cards. Their Applications to Science and Industry," New York, Reinhold Publishing Co., 1951.
(13) Pietsch, E., *Nachrichten fuer Dokumentation*, **3,** No. 1 (1952). Mechanisierte Dokumentation -ihre Bedeutung fuer die Oekonomie der geistigen Arbeit.

RECEIVED May 25, 1953.

QD
1
A355
#8

Book Bindery
West Spfld., Mass.